NEW DIMENSIONS OF AGRICULTURAL MARKETING

New Dimensions of Agricultural Marketing

Mahesh Khurana

CYBER TECH PUBLICATIONS
4264/3, Ansari Road, Daryaganj, New Delhi-110002 (India)
Ph.: 011-23244078. Fax: 011-23280028
E-Mail: cyberpublicationsdelhi@yahoo.com

NEW DIMENSIONS OF AGRICULTURAL MARKETING

Mahesh Khurana

First Edition 2009

Published by :

G.S. Rawat for Cyber Tech Publications
4264/3, Ansari Road, Daryaganj, New Delhi-110002 (India)
Ph.: 011-23244078. Fax: 011-23280028
E-Mail: cyberpublicationsdelhi@yahoo.com

Printed at : **Amit Enterprises**, Maujpur, Delhi, (9810878184)

Preface

The farm mechanization has been well received world over as one of the important elements of modernization of agriculture. In India, though there has been a considerable progress of mechanization in agriculture, its spread has been in the most uneven manner. Hitherto, governmental efforts have mostly been confined to the promotion of manual and animal-drawn tools and implements. Power-drawn equipment have also gained momentum due to the concerted efforts of the Government, Credit Institutions and the Industry. However, efforts for the identification of specific farm implements and machines, for different agro climatic zones, as well as their promotion in the respective zones seem to be lacking. This has been a constraint at the all-round and even spread of development of agriculture in the country.

Mechanization of agriculture should not only be guided by the goal of higher returns to the farmers and to the Industry, but also by its contribution to the balanced agricultural development of the different regions / areas having diverse socioeconomic and agro-climatic conditions. Thus, considering the expanding needs of the agricultural mechanization in the country, it has become necessary to ensure that the requirements of the farmers are duly met in terms of the local needs, quantity and quality of inputs and social benefits. This can only be accomplished by taking definite policy measures and the strategic planning. Keeping abreast with the recent advances & development in agricultural mechanization taking place all around, upgradation of technology is needed to be a continuous process.

—*Author*

CONTENTS

1

INDIA: AGRICULTURE AND RURAL DEVELOPMENT

A Saga of Success

From a nation dependent on food imports to feed its population, India today is not only self-sufficient in grain production but also has a substantial reserve. The progress made by agriculture in the last four decades has been one of the biggest success stories of free India. Agriculture and allied activities constitute the single largest contributor to the Gross Domestic Product, almost 33 percent of it. Agriculture is the means of livelihood of about two-thirds of the workforce in the country.

This increase in agricultural production has been brought about by bringing additional area under cultivation, extension of irrigation facilities, the use of improved high-yielding variety of seeds, better techniques evolved through agricultural research, water management, and plant protection through judicious use of fertilisers, pesticides and cropping practices.

Crops

The 1970s saw a multi-fold increase in wheat production that heralded the Green Revolution. In the next decade rice production rose significantly; in 1995-96, rice production was 79.6 million tonnes. The total grain production crossed 211 million tonnes in 2001-02, a big leap from 51 million tonnes in 1950-51.

To carry improved technologies to farmers, a National Pulse Development Programme, covering 13 states, was launched in 1986. Efforts to boost pulse production were augmented further by the Special Food Production Programme. In 2001-02, pulse production was 13.52 million tonnes. With some states offering more than the statutory minimum price, sugar cane production also received a boost, and in 2001-02 a record 292.2 million tonnes was registered.

Irrigation

As efforts continued to increase the irrigation potential in the country, the last 40 years saw the gross irrigated area reach 85 million hectares. Flood forecasting has become an important activity over the years. Over 500 hydrological stations collect and transmit data through 400 wireless stations for issuing forecasts for 157 sites. About 5,000 forecasts are issued in a year with 94 percent accuracy. The country also receives international support, with the World Bank as a primary source, for developing its water resources. International cooperation is also envisaged in setting up a National Centre for Information on Water and Power. As there is a broad seismic belt in the country, particularly along the Himalayan, and the Kutch region and parts of Maharashtra, a scheme is being evolved to collect all data on seismic activity at various dam sites.

Fertilisers

The fertiliser industry in India has grown tremendously in the last 30 years. The Government is keen to see that fertiliser reaches the farmers in the remote and hilly areas. It has been decided to decontrol the prices, distribution and movement of phosphatic and potassic fertilisers. Steps have been taken to ensure an increase in the supply of non-chemical fertilisers at reasonable prices. There are 66 fertiliser quality control laboratories in the country. Since biofertilisers are regarded as an effective, cheap and renewable supplement to chemical fertilisers, the Government is implementing a National Project on Development and Use of Bio-fertilisers. Under this scheme, one national and six regional centres for organising training, demonstrating programmes and quality testing of bio-fertilisers has been taken up.

It was a challenging decision of the Government to take Bombay High gas through a 1,700 km pipeline to feed fertiliser plants located in the consumption centres of North India. However, the major policy which has ensured the growth of the fertiliser industry is the thrust on accelerating fertiliser consumption by fixing, on the one hand, low and uniform price for fertilisers, and on the other hand providing the manufacturers adequate compensation through the retention price and subsidy scheme. As expected, fertiliser nutrient demand has gone up from 0.29 million tonnes in 1960-61 to 16.7 million tonnes at the end of 2000-01, compared to 12.15 million tonnes during 1992-93.

Fisheries

Fish production achieved an all-time high of 5.6 million tonnes at the end of 2001-02. Programmes that have helped boost production include the National Programme of Developing Fish Seeds, Fish Farmers' Development Agencies and Brackish Water Fish Farmers' Development Agencies. The Central Institute of Fisheries Nautical and Engineering Training trains the necessary manpower. To diversify fishing methods and introduce processed fish products on a semi-commercial scale, an Integrated Fisheries Project has been launched. A National Fisheries Advisory Board has also been established.

FOOD PROCESSING

The Ministry of Food Processing Industries, set up in July 1988, is the central agency of the Government responsible for developing a strong and vibrant food-processing sector with a view to create increased job opportunities in rural areas, enable the farmers to reap benefit from modern technology, create surplus for exports and stimulating demand for processed food. A new seeds policy has been adopted to provide access to high-quality seeds and plant material for vegetables, fruit, flowers, oilseeds and pulses, without in any way compromising quarantine conditions. Initiatives have been taken to encourage private sector investment in the food processing industry.

AGRICULTURAL RESEARCH

The apex body for education, research, training and transfer of technology in the field of agriculture is the Indian Council of Agricultural Research (ICAR), established in 1929. India's transformation from a food deficit to a food surplus country is largely due to ICAR's smooth and rapid transfer of farm technology from the laboratory to the land.

ICAR discharges its responsibilities through 43 research institutes, four national research bureaux, 20 national research centres, nine project directorates, 70 all-India coordinated research projects, and 109 Krishi Vigyan Kendras (farm science centres). Besides, the programme of Agricultural Education is coordinated by ICAR with the curricula and other normative guidance given to the 26 agricultural universities and four national research institutes.

OILSEEDS PRODUCTION

A Technology Mission on Oilseeds was launched in 1986 to increase production of oilseeds in the country and attain self-sufficiency. Pulses were brought under the Technology Mission in 1990. After the setting up of the Technology Mission, there has been consistent improvement in the production of oilseeds. The oilseed production, which was 108.3 lakh tonnes in 1985-86, has increased to 247.30 lakh tonnes in 1998-99, the highest production so far. However, due to unfavourable weather conditions prevailing in the major oilseeds-growing states, the oilseed production during 1999-2000 and 2000-01 suffered some decline.

The increase in production was largely contributed by soyabean, rapeseed and mustard. Production of pulses has seen many ups and downs, which is expected to be checked under the mission. The country grows mainly nine oilseeds, with groundnut, rapeseed and mustard accounting for 62 percent of total production. Lately, soyabean and sunflower have shown major growth potential.

DRINKING WATER

A Technology Mission on Drinking Water and Related Water Management has been constituted to cover the residual problem villages and provide potable water at 40 litres per capita per day, and 70 litres per capita per day in desert areas inclusive of 30 litres for cattle. The mission is tackling the problem through 55 mini-missions in project districts and countrywide problem-oriented sub-missions. A Village Level Operation and Maintenance (VLOM) pump called India Mark-II has been developed and is being exported to 40 countries.

AGRICULTURE IN INDIA

Agriculture in India is one of the most prominent sectors in its economy. Agriculture and allied sectors like forestry, logging and fishing accounted for 18.6% of the GDP in 2005 and employed 60% of the country's population. It accounts for 8.56% of India's exports. About 43% of India's geographical area is used for agricultural activity.

Despite a steady decline of its share in the GDP, agriculture is still the largest economic sector and plays a significant role in the overall socio-economic development of India. The monsoons play a critical role in the Indian sub-continent's agriculture in determining whether the harvest will be bountiful, average, or poor in any given year. The entire rainfall in the sub-continent is concentrated in the few monsoon months.

Agriculture in India is constitutionally the responsibility of the states rather than the central government. The central government's role is in formulating policy and providing financial resources for agriculture.

The government administers prices of essential commodities to protect farmer's interests. It also administers other commodities which are produced by government-controlled companies, like petroleum, coal, nitrogenous fertilizers, etc. Other than these, most agricultural commodity markets operate under the normal forces of demand and supply.

HISTORY OF INDIAN AGRICULTURE

Evidence of the presence of wheat and some legumes in the 6th millennium BC have been found in the Indus Valley. Oranges were cultivated in the same millennium. The crops grown in the valley around 4000 BC were typically wheat, peas, sesame seed, barley, dates and mangoes. By 3500 BC cotton growing and cotton textiles were quite advanced in the valley.

By 3000 BC farming of rice had started. Other monsoon crops of importance of the time was cane sugar. By 2500 BC, rice was an important component of the staple diet in

Mohenjodaro near the Arabian Sea. The Indus Plain had rich alluvial deposits which came down the Indus River in annual floods. This helped sustain farming that formed basis of the Indus Valley Civilization at Harappa. The people built dams and drainage systems for the crops. By 2000 BC tea, bananas and apples were being cultivated in India. There was coconut trade with East Africa in 200 AD. By 500 AD, egg plants were being cultivated.

Agricultural Movements

Green Revolution

After independence, considering India's growing population, the government took steps to increase the food production. Yields per unit area of all crops have grown since 1950. The 1970s saw a huge increase in India's wheat production. This is known as the Indian Green Revolution. Reasons for the growth are the special emphasis placed on agriculture and steady improvements in irrigation, technology, application of modern agricultural practices and provision of agricultural credit and subsidies

Operation Flood

Operation Flood was the name of a rural development programme started by the National Dairy Development Board (NDDB) in 1970 with the objective of creating a nationwide milk grid. This movement followed the Indian green revolution and helped in alleviating poverty and famine levels from dangerous proportions in India during the era. It resulted in India becoming the largest producer of milk and milk products, so it is also called the White Revolution of India.

Production

India is ranked second in the world in terms of agricultural output. India is the largest producer in the world of milk, cashew nuts, coconuts, tea, ginger, turmeric and black pepper. It also has the world's largest cattle population (193 million). India ranks second worldwide in farm output.

It is the second largest producer of wheat, rice, sugar, groundnut and inland fish. It is the third largest producer of tobacco. India accounts for 10 per cent of the world fruit production with first rank in the production of banana and sapota. Given below is a chart of trend of output of cereals and major foodgrains as published by the Department of Food and Public Distribution with figures in tonnes.

Year	Cereals	Rice	Wheat	Coarsegrains	Pulses
2001–02	199,480,000	93,340,000	72,770,000	33,370,000	13,370,000
2004–05	192,730,000	87,800,000	73,030,000	31,880,000	13,670,000

Despite high growth, international comparisons reveal that the average yield in India is generally 30% to 50% of the highest average yield in the world.

Productivity

The following table shows the harvest per acre of farmland in 2007 with figures in tonnes.

Commodity	Output per acre
National Average	0.29
Sugarcane	26.15
Potato	6.32
Wheat	0.84
Rice	0.70
Corn	0.64
Groundnut	0.37
Soyabean	0.32

The low productivity in India is a result of the following factors:

- Illiteracy, reforms and inadequate or inefficient finance and marketing services for farm produce.
- The average size of land holdings is very small (less than 20,000 m^2) and are subject to fragmentation, due to land ceiling acts and in some cases, family disputes. Such small holdings are often over-manned, resulting in disguised unemployment and low productivity of labour.
- Adoption of modern agricultural practices and use of technology is inadequate, hampered by ignorance of such practices, high costs and impracticality in the case of small land holdings.
- Irrigation facilities are inadequate, as revealed by the fact that only 53.6% of the land was irrigated in 2000-01 which result in farmers still being dependent on rainfall, specifically the Monsoon season. A good monsoon results in a robust growth for the economy as a whole, while a poor monsoon leads to a sluggish growth. Farm credit is regulated by NABARD, which is the statutory apex agent for rural development in the subcontinent.

In the last few decades several farmers have committed suicide especially in the states of Andhra Pradesh, Maharashtra, Karnataka, and Kerala. Combating this has become a major challenge for these governments. Some of the causes for the deaths include indebtedness of small and marginal farmers and repeated crop failures.

Waste

As per the report of the 11th Planning Commission, preventable post-harvest losses of foodgrains in India were estimated at about 20 million tonnes a year, which is nearly 10.5 percent of the total production due to inadequate storage, security and distribution facilities.

AGRICULTURE IN THE INDIAN STATES

Minor crop Areas in India:

P Pulses

s Sugarcane

J Jute

Cn Coconut

C Cotton

T Tea

About 70% population of the India is directly dependent on agriculture, also agriculture is the main component for most of the state economies in India. The Punjab region is suited for wheat-growing. The south Indian states of Andhra Pradesh, Tamil Nadu and Karnataka are also major producers of rice. These states are mostly irrigated by the rivers Krishna, Godavari and Cauvery. Haryana is self-sufficient in food production and the second largest contributor to India's central pool of food grains. Haryana contributed significantly to the Green Revolution in India in the 1970s. The high-altitude states of Himachal Pradesh and Jammu and Kashmir are ideal for production of apples. Arunachal Pradesh also has a large number of fruit orchards. Tea is the other produce of the high altitude regions of Assam, West Bengal (Darjeeling), Tripura, Tamil Nadu (Ooty), Himachal Pradesh and Kerala. Assam produces some of the finest and most expensive teas in the world (see Assam tea). However Assam tea is grown at elevations near sea level, giving it a malty sweetness and an earthy flavor, as opposed to the more floral aroma of highland (e.g., Darjeeling, Taiwanese)

In the north-eastern states like Arunachal Pradesh and Nagaland, shifting cultivation known locally as jhum, was practised by the tribal groups, but that has come to be less practised. In mountainous states like Sikkim, farming is done on terraced slopes. The state has the highest production and largest cultivated area of cardamom in India. Karnataka is the largest producer of coffee in India and accounts for 59% of the country's coffee production. It is grown mostly in the lower slopes of the Western Ghats in Kodagu district. Karnataka is also the largest producer of sandalwood based products like perfumes. Sandalwood comes from the dense forested areas of southern Karnataka. The state is also the largest producer of raw silk and Mysore silk saris are famous. The other state manufacturing silk is Tamil Nadu. Rajasthan is among the largest producers of edible oils in India and the second largest producer of oilseeds. The state is also the biggest wool-producing state in India. Kerala produces 96% of national output of pepper. It is also the largest producer of spices which include cardamom, vanilla, cinnamon, and nutmeg. Kerala is also the largest producer of natural rubber in India (91%). The other state producing rubber is Tripura.' 30 percent of Mizoram is covered with wild bamboo forests, and accounts for 40 percent of India's 80-million-ton annual bamboo crop. There has been an economic shift towards agriculture for some states like Bihar after the mineral rich areas were carved out to form the new state of Jharkhand.

Research

The Indian Agricultural Research Institute (IARI) established in 1905, was responsible for the research leading to the "Green revolution" of the 1970s. The Indian Council of Agricultural Research (ICAR) is the apex body in agriculture and related allied fields, including research and education. The Union Minister of Agriculture is the President of the ICAR. The Indian Agricultural Statistics Research

Institute develops new techniques for the design of agricultural experiments, analyses data in agriculture, and specializes in statistical techniques for animal and plant breeding.

ALL INDIA PRODUCTION OF FOODGRAINS (JULY-JUNE)				
			(IN MILLION TONNES)	
GRAINS	**2001-2002***	**2002-2003***	**2003-2004@**	**2004-2005***
		(FINAL)		
RICE				
KHARIF	80.52	63.08	78.24	73.29
RABI	12.82	8.74	10.04	14.51
TOTAL	93.34	71.82	88.28	87.80
WHEAT				
RABI	72.77	65.76	72.11	73.03
COARSEGRAINS				
KHARIF	26.71	19.99	32.37	24.60
RABI	6.66	6.08	5.75	7.28
TOTAL	33.37	26.07	38.12	31.88
TOTAL CEREALS				
KHARIF	107.23	83.07	110.61	97.89
RABI	92.25	80.57	87.90	94.84
TOTAL	199.48	163.64	198.51	192.73
PULSES				
KHARIF	4.84	4.15	6.16	5.04
RABI	8.53	6.97	8.78	8.63
TOTAL	13.37	11.12	14.94	13.67
TOTAL FOODGRAINS				
KHARIF	112.07	87.22	116.77	102.93
RABI	100.78	87.55	96.68	103.45
TOTAL	**212.85**	**174.77**	**213.45**	**206.38**
(SOURCE : DIRECTORATE OF E & S) @ 2nd Advance Estimates as on 5.8.04 (*) First Advance Estimates as on 19-01-2005				

India is now trying to adopt organic farming, but as per the survey conducted in the year 2006-2007, the total production and the total requirement of organic manures and biofertilizers in India was estimated approximately to be 12.2 million tonnes and 15.7 million tonnes respectively. Hence, the research is now focussing towards minimizing the gap between the supply and the demand of organic manures and biofertilizers. Prof. M.S. Swaminathan is known as "Father of the Green Revolution" and heads the MS Swaminathan Research Foundation. He is known for his advocacy of environmentally sustainable agriculture and sustainable food security.

AGRICULTURE AND FOOD SECURITY

Pre-Independence India suffered repeated famines, drought and food shortages. But following the Green Revolution in the '60s, yields and foodstocks rose manifold. Now, 30 years later, Indian farmers have realised the follies of their tryst with intensive agriculture. Despite 70 per cent of the population being engaged in agriculture and allied activities, declining foodgrain production and access to food remain the two biggest problems confronting the country. Liberalisation has made things worse: commercial crops are eating into the fertile land tracts meant for essential foodgrains. And six years after the World Trade Organisation came into existence, the anticipated gains for India from the trade liberalisation process in agriculture are practically zero.

Agriculture Minister C Subramaniam sat huddled with his senior staff. There was tension in the air. Depending on imported foodgrains coming from 12,000 miles away, particularly when it came in driblets following US President Lyndon Johnson's directive to "teach India a lesson", was humiliating enough. What worried him was that food stocks had reached such a precarious low that they were only sufficient for another two weeks. Still worse, there was nothing in transit. Recalls Subramaniam, "As a last resort, I told my officials and experts to identify the nearest food-carrying ships on the ocean throughout the world. I said we would identify those carrying wheat to other countries and appeal to the US President to divert them to India if other countries could wait for another six to eight weeks."

This was in 1966-67, the critical year of drought when India imported 11 million tonnes of foodgrain. A year earlier, India had imported 10 million tonnes of foodgrain. Around this time, the Paddock brothers, often referred to as 'prophets of doom', concluded in their book, Famine 1975, that by the mid-'70s at least half of India would be led to a slaughter house. Considering that India had a long history of hunger and famine, and the subsequent failure of the government's effort to boost agricultural production in the post-Independence era, not many disagreed with the conclusive analysis drawn by William and Paul Paddock.

After all, for a country literally surviving on a 'ship-to-mouth' existence, there was little hope. Soon after Independence, India had begun to seek food aid and in fact emerged as the biggest food importer of this century. In 1951, India got ad hoc assistance of 2 million tonnes of foodgrain to tide over the crisis arising from severe drought in several parts of the country. This was the first major food security on a significant scale after Independence.

Before 1947, Indian history was replete with famine, drought and food shortages. Between 1770 and 1880, as many as 27 food scarcities and famines were recorded. At least 20 million lives were lost in India in about 20 famines that had struck since 1850.

Much of this loss was because of misplaced colonial policies which aimed to derive maximum economic gain at the cost of human suffering and misery. After the British had created a transport infrastructure in the first half of the 19th century, they began encouraging farmers to grow crops that could be exported.

The boom in export trade accompanied by rising prices made farmers shift to cash crops like cotton, indigo, poppy and sugarcane. The area under foodgrain subsequently shrank. In other words, efforts to improve agriculture in colonial India were directly linked to the needs of British industries.

In the first half of the 20th century, India's foreign trade showed the same colonial pattern with exports mainly comprising foodgrain and cotton, jute, oilseeds, opium and indigo. Against this, imports consisted mainly of consumer goods. Meanwhile, the cultivator's preference for commercial crops continued.

As is happening now, the trend during the pre-Independence years showed that the best lands available were used for the cultivation of commercial crops. The cultivation of foodgrains was relegated, to some extent, to inferior lands, the yield of which is much less even in normal years. The result was a steady decline in foodgrain output till the Second World War. Per capita food availability from domestic production too had declined. As B M Bhatia observes in The Famines of India: "India kept the precarious balance between increasing population and declining food production partly by turning from an exporter into a net importer of foodgrains but largely by the easy, though dangerous, expedient of pushing down the already low per capita cereal consumption to still lower levels."

The big gap between minimum requirement and supply continued throughout the post-War period. Till the 1960s, Indian agriculture was not sufficiently able to meet domestic requirements and the country had to rely on foodgrain imports.

SOWING THE SEEDS OF THE GREEN REVOLUTION

Realising the disastrous consequences of the growing gap between the rates of growth of population and food production, a vigorous 'grow more food' campaign was launched. The key goal of economic planning shifted to attaining self-sufficiency. The First Five-Year Plan, following food imports in 1951, gave maximum importance to the growth of agriculture. Among the measures initiated in the '50s to stimulate foodgrain production were land reforms, irrigation, fertiliser production, strengthening of research and the organisation of a national farm extension service. Consequently, increased harvests in the mid-1950s relieved the economy.

It was around this time that the government adopted the US Land Grant model of agricultural research. Agricultural universities, beginning with the Gobind Ballabh Pant University of Agriculture and Technology in Pantnagar in Uttar Pradesh, were gradually set up in all the states. The Land Grant model entailed linking research with extension, and taking research from the laboratories to the farms through an extension network linked with the university. In 1961, the government launched the Intensive Agriculture District Programme (IADP) to enhance productivity in the irrigated areas. Although the aim was to introduce improved seed along with a package of agronomic practices to take advantage of the available irrigation, the programme did not come up to expectations for want of high-yielding varieties of wheat.

The tall Indian wheat varieties did not provide an economic response to the application of chemical fertilisers and irrigation water, which agricultural scientists were desperately looking for. These had a tendency to fall down or lodge when fertilisers were applied, as a result of which the yields stagnated at less than one tonne per hectare.

Even though breeding for improved non-lodging wheat varieties was initiated in the '50s, it did not meet with much success.

The missing link was provided by Dr Norman Borlaug, who was then working with CIMMYT in Mexico. A few dwarf wheat strains of spring wheat that he sent to the Indian Agricultural Research Institute in New Delhi, looked impressive. In the next two to three years, a wide range of dwarf material was tested under the All India Wheat Research Project. In August 1964, the then Professor and Head of Department of Plant Breeding and Genetics at the IARI, Dr M S Swaminathan, who later emerged as the architect of India's Green Revolution, proposed the launching of a National Demonstration Programme in farmers' fields. This would verify the results obtained in research plots and introduce farmers to new opportunities opened up by semi-dwarf varieties for considerably improving the productivity.

Says Swaminathan, "When small farmers, who with the help of scientists organised the National Demonstration Programme, harvested over 5 tonnes of wheat per hectare, its impact on the minds of other farmers was electric. The clamour for seeds began and the area under high-yielding varieties rose from 4 hectares in 1963-64 to over 4 million hectares in 1971-72. A small government programme became a mass movement." Subsequently, the government imported 18,000 tonnes of semi-dwarf wheat from Mexico in 1966 to be distributed in time for immediate sowing by farmers in the northwestern regions.

The seeds of the Green Revolution were, therefore, truly sown in the mid-'60s. What followed next is already part of contemporary history. Wheat production rose to 17 million tonnes in 1968, an increase of 5 million tonnes over the highest of 12 million tonnes harvested in 1964. Indira Gandhi officially recorded the impressive strides in agriculture by releasing a special stamp entitled 'Wheat Revolution' in July 1968. The success of wheat was later replicated in rice. Productivity increases were also recorded in cotton, sugarcane, millets and oilseeds.

MEASURES TO AVOID FAMINE

The famine-avoidance strategy that India launched had all the necessary ingredients for a sustained effort. In 1965, the Agricultural Prices Commission was constituted, with the basic objective of assuring fair prices for farm produce. The Commission, an autonomous body, works out the cost of cultivation for 22 agricultural commodities, and suggests a price to the government that includes profit that should serve as an incentive to grow more. The administered price of wheat, for instance, was raised by 71 per cent between 1972 and 1981. Similarly, the price of rice was jacked up by 81 per cent during the same period. Later, to sustain the tempo of the rice revolution, the relative price structure was moved in its favour. In the early phase of the Green Revolution, the price policy favoured wheat, but after the mid-'70s it tilted in favour of rice. For oilseeds and pulses, the reverse was put into action. Relatively high support prices were announced in the early-'90s to make their cultivation profitable.

Perhaps the most significant aspect of the price support mechanism has been the insulation of farmers against a decline in prices. Wheat production was very high in 1971-72, as a consequence of which prices should have declined by at least 25 per cent. Since the government purchased all the wheat that was offered, farmers were actually paid 7 per cent more. Again, in 1982-83, when wheat production was 14.3 per cent higher than the previous year, the prices should have actually declined by 32.5 per cent. With the government purchasing all the wheat that flowed into the markets, however, the farmers realised 18 per cent more.

The biggest achievement of the Indian food policy, dependent more and more on buffer stocking and operational stock holding, has been the avoidance of famine-like conditions. It was with the basic objective of curbing consumption and ensuring an equitable distribution of available food supplies, especially in the deficit areas and among the poorer strata of society, that the Public Distribution System (PDS) was introduced. PDS supplies were also used for the 'food for work' programme as well as other anti-poverty programmes. At present, PDS covers more than 80 million families and the total foodgrains distributed through a network of 40,000 fair price shops accounts for over one-third of the total trade in foodgrain.

AIMING AT SELF-SUFFICIENCY

The Green Revolution was certainly not a stroke of luck. The resilience was the result of progressive farm policies, determination, sustained efforts of scientists and the vigour of the Indian farmers who readily adopted the improved technology. This is what the Rockefeller Foundation had to say about the magnificent strides taken: " ... The speed with which India's farmers and scientists suddenly gave their country an approach to an abundant food supply, has never been duplicated on an equal scale anywhere in the world, including the scientifically sophisticated United States."

Since the early-'70s, the country has avoided famine even during adverse weather conditions, through a carefully designed food security system involving the maintenance of both substantial grain reserves and an extensive PDS. The devastating drought of 1987, rated as the most severe of this century, affected nearly 155 million hectares. In spite of this, no starvation and hunger deaths were reported. This was essentially the result of excellent food management based on foodgrain reserves built largely from homegrown wheat and rice.

From the late-'60s, the rate of growth in food production generally exceeded the rate of growth of population. Despite the continuous growth in population, the per capita availability of foodgrain increased from 157 kg per year in 1955 to 177 kg per year in 1995, an increase of 15.7 per cent. Food production has risen roughly four times since the days of Independence, from 50 million tonnes in 1950-51 to 198 million tonnes in 1996-97. At the same time, the population has risen from 300 million to 960 million. Milk production too has risen to 68 million tonnes, the highest in the world. Food self-sufficiency was achieved in the late-'70s with the maintenance of a sizeable food buffer. Lately, India has begun to export foodgrain. About 5 million tonnes of rice and wheat were exported during the last few years. Skimmed milk powder amounting to about 5,000 tonnes has also been exported.

The increase in foodgrains was, however, disproportionate. The highest rate of growth was achieved in Punjab and Haryana, where, at a compound growth rate of 4.63 per cent, foodgrain production jumped from 7.23 million tonnes in 1964-65 to reach an all-time high of 30.33 million tonnes in 1995-96. Uttar Pradesh recorded a compound growth rate of 3.01 per cent. Production in Maharashtra, Madhya Pradesh, Bihar, Orissa and the northeastern states continues to stagger. Tamil Nadu and Andhra Pradesh, on the other hand, have recorded significant increases in rice yields, with production rising to 9.17 million tonnes and 11.58 million tonnes, respectively.

The Green Revolution was aimed at the better-endowed regions. For millions of farmers languishing in the drylands, constituting more than 70 per cent of the cultivable lands, it continues to be a futile struggle. Despite emphasis on dryland farming during the past several decades, the scenario still remains grim. The undulating topography and the irregular rainfall pattern have combined to aggravate the situation. That the drylands produce about 42 per cent of the country's food shows that the future of farming lies in these areas. Nearly 83 per cent of sorghum, 81 per cent of pulses and 90 per cent of oilseeds grown in the country come from these areas. The poor yields and the fluctuations in production are indications of the scant attention drylands have received from policymakers and planners.

The problem of increasing productivity on drylands has serious socio-economic implications. With every passing year, the gap between the farmer's yields in irrigated areas and in the dry farming region is widening. One year of drought is enough to push a farmer into a deep well of poverty for another two to three years. Drought is a recurring phenomenon in arid and semi-arid areas. Fifty years after Independence, life for millions of people somehow surviving in the drylands continues to be worse than before.

DECLINE IN PRODUCTION

Thirty years after the dawn of the Green Revolution, Indian farmers are realising that their love affair with intensive agriculture is on the decline. Despite a bountiful monsoon for the tenth year in a row, harvests are not as plentiful as expected. However, with no let-up in the ever-increasing rise in population, Indian agriculture is once again at the crossroads.

Alarm bells have been ringing for quite some time. For nearly a decade, agricultural production had stagnated. The spectacular yield growth recorded in the post-Green Revolution years in Punjab and Haryana has receded into history. Of the multiple problems confronting agriculture, rapid fragmentation of land holdings is keeping pace with increasing population. In 1976-77, the average size of the holdings was estimated at 2 hectares, while in 1980-81, it came down to 1.8 hectares. Today, it stands at a mere 0.2

hectares. The total number of land holdings in 1981 were around 89 million; today these have crossed 100 million. By the turn of the century, the average land holding will come down to 0.11 hectares. It is quite obvious that with such small land holdings, Indian agriculture cannot adopt high-tech farm practices.

Basking in the afterglow of the Green Revolution, farming and agriculture have ceased to attract serious attention. In 1995-96, foodgrain production slumped to minus 3.60 per cent. In 1997-98, it further declined to minus 3.70 per cent, the worst-ever since the heady days of the Green Revolution. Foodgrain production in the frontline agricultural states of Punjab, Haryana and western Uttar Pradesh, comprising the country's food bowl, has decelerated. The miracle that began with wheat was replicated in rice. This is where the cost-benefit ratio has turned negative. The downward spiral in food production continues through the southern regions of the country. Tamil Nadu, another Green Revolution area, is under tremendous strain from intensive cultivation. In Karnataka, the negative trend in yield levels of all food crops, barring cotton and sugarcane, are all too apparent. Farming in Karnataka can be clearly separated in two distinct classes, the 'creamy layer' of corporate agriculture occupying the fertile and irrigated areas, and the remaining low productive tracts at the mercy of the rains, constituting the tiny and small land holdings. Kerala too, picturesque and verdant with tropical forests, is passing through its worst-ever crisis on the foodgrain front.

As if this is not enough, the number of landless in rural areas too has been multiplying over the past few decades, at an estimated rate of two million every year. The rapid expansion in the number of land holdings, with less than a hectare under plough, has meant an additional 6,00,000 farmers every five years or so. The negative terms of trade for agriculture and the declining public sector investments in farming are indices of the sluggishness in this sector.

The Economic Survey for 1996-97 has further sounded a warning. The compound growth rate in foodgrain production at 1.7 per cent between 1991 and 1996, is lower than the annual population growth of 1.9 per cent. The survey adds that even a marginal fall of three to four per cent in foodgrain output can cause the prices of essential commodities to escalate sharply, necessitating government intervention by way of resorting to imports. This happened in 1993 and again in 1996.

The philosophy of agricultural planning is changing. Gone are the days when the nation's emphasis was solely on attaining self-sufficiency in foodgrain production. Gone also are the days when farmers were newly independent India's heroes, revered for their role in keeping hunger and starvation at bay. Today, at a time when food production struggles to keep pace with the burgeoning population growth, farmers are being asked to diversify, produce crops that are suitable for export and to compete in the international market. With the promise of cheap food available off the shelf in the global market, the focus has shifted from agriculture to industry, trade and commerce, from the small and marginal farmers to the agri-processing companies, which alone can bring in investments and add value to produce.

The restructuring of the Indian economy too has put the emphasis on agri-exports. Cultivation of staple food is being replaced by cash crops, tomatoes in place of wheat, durum wheat (for bakery purposes) replacing wheat as a staple diet in Punjab and Haryana, flowers in place of rice, and so on. In the coastal areas, private enterprise is taking away the fish catch, depriving the local communities of a livelihood and their only source of nutrition. In Kerala, vast tracts of forests and paddy fields have been converted into rubber, coffee and coconut plantations. Every year, about 25,000 hectares of good paddy land are being diverted for non-paddy purposes. The structural transformation is not only peculiar to Kerala. It is happening in almost all the states. Commercial crops are eating into the fertile land tracts meant for growing essential foodgrains, thereby making the states rely more heavily on the PDS. The diversion of good agricultural land, which in any case is limited, to commercial farming and even industries, is further restricting the ability to grow enough foodgrain.

IN THE ERA OF WTO

Kerala has already been at the receiving end. Over the past few years, ever since economic liberalisation became the development mantra, the state has been flooded with cheap and highly subsidised agricultural imports, throwing its agrarian economy out of gear. Whether it is the import of palm oil or rubber or coffee, almost every aspect of the state's socio-economy has been negatively impacted.

Coconut prices have crashed, down from Rs 10 to Rs 2. Rubber has plummeted from Rs 60 to Rs 16 and coffee from Rs 58 in 1999 to Rs 30 per kg in 2000. Even spices have not been spared, with pepper prices falling from Rs. 2,600 to Rs. 1,300 per quintal in the consecutive period. Kerala is not alone.

The fallout from the emerging global trade paradigm is being felt all over the country, though not in the same magnitude. Before we go any further however, it is important to understand how the market rules play, and then try to understand the implications for Indian agriculture.

It is now official. Six years after the World Trade Organisation (WTO) came into existence on January 1, 1995, the anticipated gains for India from the trade liberalisation process in agriculture are practically zero. The ministry of agriculture as well as the ministry of commerce have admitted that the hopes from an international regime that talked of establishing a fair and market-oriented agricultural trading system have been belied.

It is time to draw a balance sheet of the gains and losses that accrued from the implementation of the WTO's Agreement on Agriculture, which had incorporated three broad areas of commitments from member states, namely in market access, domestic support and export subsidies.

The underlying objective of this was to correct and prevent restrictions and distortions in world agricultural markets. Six years later, it has been established that these measures have only protected the farmers and the farming systems of the developed countries. On the other hand, the trading regime has ensured that developing countries take time-bound initiatives to open up their domestic markets for cheap and highly subsidised imports of agricultural commodities.

EASIER MARKET ACCESS NEEDED

A recent study by the Food and Agricultural Organisation of the United Nations (FAO) concludes that there has been hardly any change in the volume of exports. Tariff peaks or high import duties continue to block exports from the developing countries.

Tariffs still remain very high, especially in the case of cereals, sugar and dairy products. Sanitary and phytosanitary measures, which were enforced to ensure the quality of imported products, actually continue to be a major barrier in diversifying exports in horticulture and meat products. Selective reduction in tariffs by developed countries has also blocked exports from developing countries. To top it all, only 36 countries (all developed) have the right to impose special safeguards if agricultural imports distort their domestic market!

India was forced to either phase out or eliminate the quantitative restrictions (QRs) on agricultural commodities and products by April 1, 2001. India has, therefore, opened its market. Already, cheaper imports of skimmed milk powder, edible oils, sugar, tea, arecanut, apples, coconut etc have flooded the market.

LACK OF DOMESTIC SUPPORT

Clever manipulation of their subsidy reduction commitments has in reality increased the support to farmers in developed countries. In the United States, subsidy to just 9,00,000 farmers has increased 700 times since 1996. Between 1998-2000, the US has provided an additional US $26 billion to its farmers. In absolute terms, farm support in the OECD countries increased by 8 per cent to reach the staggering figure of US $363 billion in 1998.

In India, we are being told that since our Aggregate Measure of Support (AMS), a measure of the subsidies that are provided to agriculture, are negative, we can still raise our subsidies to farmers. In reality, India is committed to doing away with agricultural subsidies under the Structural Adjustment Programme of the World Bank and IMF. In any case, India provides only US $1 billion worth of indirect subsidies to 550 million farmers!

It was anticipated that due to reduction in domestic support in developed countries, cereal production would shift from developed countries to developing countries. Empirical evidence, however, shows that such a trend is not at all visible. Moreover, with such massive subsidies intact, and with the QRs being lifted, India faces the threat of subsidised food imports.

EXPORT SUBSIDIES URGED

The agriculture ministry acknowledges that despite the rules being defined, the expected gains have eluded the developing countries. It was expected that with the removal of trade distorting measures, agricultural exports from the developing countries would increase. This did not happen. In fact, India has seen a three-fold increase in the imports of agricultural commodities and products, from about Rs 5,000 crore in 1995 to over Rs 15,000 crore in 1999-2000. In edible oils alone, the import bill has soared to Rs. 9,000 crore. The so-called fair trading system has also not helped efficient producers in realising a higher price for their products. On the contrary, prices of most agricultural commodities are declining in the world markets.

In conclusion, if declining foodgrain production and access to food remain the two biggest problems confronting the country, there must be something terribly wrong with the way we look at agriculture. With more than 70 per cent of the population still engaged in agriculture and allied activities and an equal percentage of farmers tilling an average of 0.2 hectares of land and somehow surviving against all odds, the time has come to set the balance right. Whether we accept it or not, India is gradually moving back to the pre-Green Revolution days of a 'ship-to-mouth' existence, when food was largely imported to feed the hungry. It was the political maturity of the then leadership that led to self-sufficiency on the food front. Few will still question what Jawaharlal Nehru once said: "Everything else can wait, but not agriculture."

2

SUSTAINABLE DEVELOPMENT OF RAINFED AGRICULTURE IN INDIA

INTRODUCTION

India's agricultural growth has been sufficient to move the country from severe food crises of the 1960s to aggregate food surpluses today. Underlying this growth were massive public investments in irrigation, agricultural research and extension, rural infrastructure, farm credit and rural development programs. India's agricultural sector, however, faces severe challenges for the future. Despite sizeable national food stocks (30 million tons in 1995), widespread poverty and hunger remain because agricultural and national economic growth have not adequately benefitted disadvantaged regions and the poor.

The demand for basic staples, non-food grains, and exports is increasing. At the same time, resources are shrinking and the productivity of some resources already being utilized is threatened by environmental degradation. Growth in total factor productivity is reported to have declined slightly in major crops. Returns to investment in agricultural research and rural infrastructure are reported to be high, but these investments remain low.

Most of the increase in agricultural output over the years has taken place under irrigated conditions. The opportunities for continued expansion of irrigated area are limited, however, so Indian planners increasingly are looking to rainfed, or unirrigated agriculture to help meet the rising demand for food projected over the next several decades. Despite the historic bias in favor of irrigated agriculture in terms of research and infrastructural investments, rainfed agriculture has always been an important part of the agricultural sector.

Rainfed agriculture accounts for about two-thirds of total cropped area (Government of India 1994b, nearly half of the total value of agricultural output. Nearly half of all food grains are grown under rainfed conditions, and hundreds of millions of poor rural people depend on rainfed agriculture as the primary source of their livelihoods.

Rainfed areas are highly diverse, ranging from resource-rich areas with good agricultural potential to resource-poor areas with much more restricted potential. Some resource-rich rainfed areas potentially are highly productive and already have experienced widespread adoption of improved seeds. In drier, less favorable areas, on the other hand, productivity growth has lagged behind, and there is widespread poverty and degradation of natural resources. Even given that rainfed agriculture should receive greater emphasis in public investments, a key issue is how much investment should be allocated among different types of rainfed agriculture. Outmigration and income diversification into the nonagricultural sector must provide the long term solution to economic development of many resource poor areas, but these opportunities currently are inadequate in relation to population growth to provide short to medium term solutions. Agricultural growth in these areas will be essential for reducing poverty and environmental problems in the decades ahead.

There is a need to identify the opportunities for stimulating agricultural growth and reducing poverty and environmental degradation in rainfed areas. Likewise, there is a need to assess the opportunity costs of diverting scarce public resources from resource-rich to resource-poor areas. The tradeoffs between investing in resource-rich and resource-poor areas in terms of their productivity, poverty and environmental outcomes need to be understood in order to guide public policy decisions toward productive outcomes.

Developing strategies for rainfed areas is difficult because of their diversity in terms of agroecological characteristics, infrastructural development, and other socioeconomic variables. On an all-India scale, for example, rainfed systems include high-rainfall agriculture in the east and northeast as well as the drought-prone areas of the Deccan Plateau. Other agroclimatic characteristics such as soil types also vary, as do infrastructure development, human capital, and other socioeconomic factors.

Across villages within a district, for example, there is wide variation in access to paved roads and public transportation to market centers. Similar diversity of agricultural systems is found even at the local level. Individual villages in the semi-arid regions, for example, often contain numerous soil types with widely differing crop production potential (Dvorak 1988). Irrigation wells are found in practically every village, so irrigated and rainfed agriculture co-exist almost everywhere.

Diversity at both the national and local scale has implications for agricultural development strategies. At the national scale, there is a need to distinguish among regions according to their constraints to agricultural development. This requires creating a typology of rainfed agriculture that would incorporate both agroecological and socioeconomic variables in order to serve as a tool for planning agricultural research and other public investments. Local-level diversity of rainfed agricultural systems, meanwhile, implies that planners must recognize that changes in policy, technology or infrastructure may have varying impacts across small areas, and that there is a limit to the extent to which external interventions can induce finely-tuned responses. Regional or district-level planning must be complemented by local initiatives that can be more responsive to specific needs.

OBJECTIVES OF THE PAPER

The overall goal of this paper is to review the important issues in rainfed agricultural development and report on the progress made in India to date. This will serve as a precursor to a detailed study to be carried out by the Indian Council of Agricultural Research, IFPRI, ICRISAT and the World Bank. That study will result in recommendations for designing a strategy to develop rainfed agriculture in India. In this paper, we compare the past performance of rainfed and irrigated agriculture and of different types of rainfed agriculture, including relatively high- and low-potential areas. We attempt to identify the factors that determine differences in performance, and we examine the possibilities for influencing those factors.

Where information is not available, we recommend further analysis that may be required as a prerequisite to formulating a thorough strategy for rainfed agricultural development. We approach the problem by reviewing the relevant literature on the subject and conducting a statistical analysis of all-India district-level data.

The database contains several agroclimatic and socioeconomic variables that we hypothesize to influence agricultural performance at the district level. The district-level approach, of course, does not permit analysis of the implications of micro-level diversity of rainfed agricultural systems.

Therefore we focus on broader indicators with implications for area-wide development efforts. In addition to the district level data analysis, we review the literature on rainfed agricultural performance in terms of technology adoption and performance, yield levels and their variability, natural resource sustainability, and poverty alleviation. We examine the role of economic and social policies, area development programs and infrastructural investments in promoting sustainable rainfed agricultural development. On some topics sufficient evidence is available to draw conclusions and make policy recommendations, and on others, additional analysis is recommended.

CHARACTERISTICS OF RAINFED AGRICULTURE

In this section we introduce some characteristics of rainfed agriculture that will influence our approach to the problem of deriving recommendations to stimulate rainfed agricultural growth.

As mentioned in the previous section, rainfed and irrigated agriculture coexist in practically every village in India. Public investment programs, however, usually cannot be targeted so precisely. For practical purposes, they need to be planned and implemented on a larger scale, such as at the village, taluk, district or state level. For example, public programs that provide credit, employ people, or build roads cannot target their efforts to either rainfed of irrigated agriculture; they can only target areas that are relatively more irrigated or more rainfed.

Price policies, on the other hand, can attempt to target rainfed or irrigated agriculture within a given location by targeting crops that may be more likely to be rainfed or irrigated. But few crops are either 100% irrigated or 100%

dryland, so some spillover will always remain. Also, every crop is grown over a large geographic area, so it is difficult to isolate the socioeconomic and agroclimatic variables affecting their performance. And while price policies are important, they are not the only approach through which policy makers can influence agricultural development.

DEFINING RAINFED AGRICULTURE

In the present investigation we use the district as the unit of analysis. We do so for two main reasons. First, through a district level focus our analysis may be relevant for public investment programs that provide infrastructure or other social services to particular areas. Second, a district focus enables us to examine the contributions of both socioeconomic and agroclimatic variables on a nation-wide basis.

The district is the smallest administrative unit for which the required data are available. To arrive at a district-level definition of rainfed agriculture, we consider the percentage of each district that is irrigated or rainfed, and consider predominantly rainfed districts as "rainfed" and predominantly irrigated districts as "irrigated." Obviously there is a certain degree of arbitrariness to any threshold we may choose to distinguish between irrigated and rainfed districts. Several previous studies have faced this same dilemma in categorizing rainfed areas. Some of them have distinguished between irrigated and rainfed districts according to certain criteria such as the amount of rainfall and the level of irrigation. Some of these studies and their definitions of rainfed areas are listed in Table 2.1.

As mentioned above, all of these definitions suffer from the inability to distinguish between rainfed and irrigated agriculture within districts, but we accept this as an inevitable limitation. Another problem is that both the rainfall and irrigation thresholds are defined somewhat arbitrarily. We discuss rainfall and irrigation thresholds in turn.

Rainfall Criteria

Among the definitions listed in table 2.1, Bapna et al, (1984) and subsequently Jodha (1985) used broad rainfall thresholds, which is important in order not to be too exclusive. At the same time, the 500-1500 mm range maintains a degree of homogeneity in the types of agriculture under analysis by excluding both very dry, desert areas and very high rainfall areas. Such areas may face unique constraints that limit comparability to agriculture under the more moderate conditions that predominate in most of the country.

Shah and Sah (1993) and Thorat (1993) use narrow rainfall thresholds with a relatively low maximum because they intended to focus on very dry (but not quite desert) areas. It is important to note that the impact of the level of rainfall on crop production is conditioned by both the distribution of rainfall over the course of the season and the factors that determine moisture retention in a given location.1 As a result, narrow rainfall thresholds such as those used by Thorat (1993) are likely to combine some areas with disparate moisture regimes and separate others with similar moisture regimes.

A narrow rainfall range probably makes sense only if limited to relatively uniform soil types. Incorporating soil types into the definition, however, introduces yet another variable and makes the definition somewhat clumsy.1 For this reason, our preference is to utilize a relatively broad range of rainfall levels such as used by Bapna et al, (1984), excluding only those that are either desert environments or extremely humid. We choose a range of 450-1600 mm average annual rainfall as the range for predominantly rainfed districts. The lower bound of 450 mm excludes the desert districts of western Rajasthan as well as one district each in Punjab, Haryana and Gujarat. The 1600 mm upper bound excludes the Himalayas, the northeastern states, Kerala, all the coastal districts of Karnataka and Maharashtra, and one coastal district each in Tamil Nadu and Gujarat.

Table 2.1 Alternative criteria to define rainfed agriculture

Authors	Criteria used
Bapnal et al (1981)	Percentage of gross cropped area under irrigation (less than 25 percent) and average annual rainfall (between 500 and 1500 mm)
Rangaswamy (1981)	Percentage of gross cropped area under irrigation (less than 30 percent) and average annual rainfall (between 375 and 1125 mm)
Jodha (1985)	Percentage of gross cropped area under irrigation (less than 25 percent) and average annual rainfall (between 500 mm and 1500 mm)
Subbarao (1985)	Percentage of gross cropped area under irrigation (less than 25 percent) and average annual rainfall (less than 970 mm)
Shah and Sah (1993)	Percentage of gross cropped area under irrigation (less than 25 percent) and average annual rainfall (between 400 and 750 mm)
Thorat (1993)	Percentage of gross cropped area under irrigation (less than 10 percent) and average annual rainfall (between 375 and 750 mm) (high intensity dry farming area)

Where more disaggregated analysis is required to examine the performance of relatively moist or dry rainfed areas, we can subdivide the rainfall criteria into a low rainfall area (<750 mm per annum), a medium rainfall area

(750-1125 mm), and a high rainfall area (>1125 mm), sometimes described as the arid, semi-arid and humid areas. To repeat the earlier caveat, these broad rainfall classes are heavily conditioned by the factors that determine moisture retention, particularly soil type.

Irrigated Area Criteria

Classifying districts by irrigated area is difficult for several reasons. First, any threshold percentage area irrigated must be defined somewhat arbitrarily, and second, in most districts irrigated area has increased steadily during the period under study.

As a result, we consider some alternate approaches to categorizing districts by irrigated area. All of the studies listed in table 2.1 use a single irrigated area threshold to distinguish between irrigated and rainfed districts.

In these studies the threshold ranges from 10 percent to 30 percent, with most defining rainfed districts as those with less than 25 percent irrigated area. 25 percent is the mean irrigated area for the years 1956-90, so according to this definition, rainfed areas are those with less than average area irrigated, while irrigated areas are those with more than average area irrigated. A number of arguments can be made about whether the figure of 25 percent is appropriate, but ultimately any definition based on such a threshold suffers from the problem that slight differences in irrigation levels will move some districts from one category to the other.

One approach is to use three categories of irrigation instead of two in order to more clearly identify the characteristics of lightly and heavily irrigated districts. This approach however, also has weaknesses, because more categories means more thresholds, which means that fewer districts will remain in one category over the entire period.

As a result, when the three irrigation categories are used as described above, nearly half of all districts shift categories at some point in the period. When only two irrigation categories are used, on the other hand, only about one quarter of the districts shift categories.

For that reason, we also conduct the analysis with only two irrigation categories, with districts with less than 25 percent area irrigated considered unirrigated, and districts with 25 percent or more area irrigated considered irrigated. We conduct the analysis twice; in one case all districts are analyzed, and in the other only districts that remain in one category or the other throughout the study period are used.

Before continuing, we briefly discuss two other criteria for defining rainfed areas that we considered but rejected. First, districts could also be subdivided by the extent of different types of irrigation, in particular, canals, wells, or tanks. The justification for this concerns the quality of irrigation services delivered to each farm. Well irrigation, for example, is controlled by the individual farmer (to the extent that the aquifer yields water), whereas under canal irrigation farmers depend more on the amount of water taken by their upstream neighbors, so they incur a greater risk of drought.

As a result, farmers with well irrigation apply more inputs and have much higher yields on average (Shah, 1993). The district-level data cover gross cropped area as opposed to net cropped area, so we control for variations in the quantity of irrigation water delivered to the farm. However, the data do not control for variations in quality or differences in farmers' response to the different levels of risk under each irrigation source. When the circumstances warrant it we may examine irrigation by source, but mainly we do not, since our main focus is on rainfed agriculture, not distinctions in irrigated types.

A second alternative criterion for defining irrigation levels would be to distinguish districts by the proportion of farmers who have access to some irrigation. In many areas, water markets or shared irrigation wells enable farmers to gain access to irrigation even if they do not own a well or are not directly serviced by a canal or a tank. As a result, the number of farmers with access to irrigation can be much larger than the number who own wells (Shah 1993).

The distinction between the proportion of area irrigated and the proportion of farmers is important because it can affect the way in which most farmers manage their crops. If a larger proportion of farmers have access to a small amount of irrigation, they may concentrate their managerial and other inputs on irrigated plots, which they may perceive to be more productive and less prone to risk of crop failure than dry plots.

In this case, perhaps dryland crops would be less productive (though more farmers will be better off). Farmers without access to irrigation, on the other hand, may devote relatively more resources to rainfed crops than those with some irrigation. Unfortunately we do not have access to district-level data on the proportion of farmers with access to irrigation, and we do not know if the proportion of farmers with access to irrigation and the proportion of area irrigated vary independently of each other across districts. Therefore we cannot consider using this definition; we raise it only to draw attention to some of the issues to consider in developing a definition of a rainfed district.

THE NEED FOR A TYPOLOGY OF RAINFED AGRICULTURE

The definition of rainfed agriculture presented in the previous section will assist us in comparing the performance of predominantly rainfed and irrigated districts. We also wish to compare the performance within rainfed agricultural districts and relate differences to a range of constraints to agricultural development.

This in turn will be useful for prioritizing and organizing agricultural research, public investment, and policy and institutional reform.

To characterize districts according to the various agroclimatic and socioeconomic variables that constrain agricultural development, we will need to construct a typology based on those variables. In this section we discuss in a bit more detail the typologies of Indian rainfed agriculture that already exist, the reasons why a new typology needs to be constructed, and the ways in which such a typology would be used.

Existing Typologies

Current typologies of Indian agriculture are based on agroecological zones. Recently, ICAR delineated 20 agroclimatic regions based on soils and climate (NBSS&LUP, 1992). 13 of these zones cover the area of this study; of the rest, 5 are in the Northeast, the Himalayas, and the Andaman, Nicobar and Lakshadweep Islands; one zone covers the high rainfall areas of the Western Ghats and the Arabian Sea coast; and one zone covers the desert in the western parts of Rajasthan and Gujarat. Figure 2.1 displays the 20 zones, and table 2.2 presents some distinguishing features of each.

ICAR's agroecological zoning system is based on variations in rainfall, soil type and temperature. Irrigation status, however, is conspicuously absent. This is a severe limitation due to the primary importance of irrigation in determining cropping patterns and productivity in most of the country. In this analysis we overcome that shortcoming by dividing zones into primarily rainfed and primarily irrigated districts. We discuss this division further below. The 20-zone system is sufficiently disaggregated to enable it to keep problems of within-zone variation to a manageable level, and the number of zones remains small enough to be manageable for most uses.

Also, the zones can be easily reaggregated for particular purposes. Recently ICAR subdivided the 20-zone typology into a total of about 50 subzones; such a disaggregated typology may be useful for certain agricultural research purposes, but for policy analysis it is too large to be functional.

In our analysis, using rainfed and irrigated districts in the 13 zones covered in our data would yield a total of 26 categories, which becomes unmanageable. For the purposes of our district-level analysis, we modify the 20-zone system so that it is sufficiently aggregated for our purposes. We create a new 5-zone system in which each zone is a combination of two or more of the zones of the 20-zone system.

The new zones, shown in table 2.3, require some explanation. First, the 20-zone system does not follow district boundaries, but in our analysis districts must remain intact. As a result, in many cases we classify a district as lying in one zone even though part of it may actually lie in another. Districts in ICAR's Zone 18 along the Bay of Bengal coast, for example, also lie in adjoining agroclimatic zones 7,8 and 12.

As a result, zone 18 drops out of the sample. Second, by combining the zones it is inevitable that some new, aggregated zones will contain substantial within-zone diversity. We find that for the case of ICAR zone 4, it makes sense to place the portion of zone 4 that lies in the Gangetic plain in one new zone, and the part that lies in upland areas of Rajasthan and Gujarat in another.

Table 2.2 ICAR's 20 agroclimatic zones

1* Western Himalayas, cold arid ecoregion, with shallow skeletal soils and length of growing period (GP) less than 90 days

2 Western plain, Kachch and part of Kathiawar peninsula, hot arid ecoregion, with desert and saline soils and GP < 90 days

3 Deccan Plateau, hot arid ecoregion, with red and black soils and GP < 90 days

4 Northern plain and central highlands including Aravalli hills, hot semi-arid ecoregion, with alluvium derived soils and GP 90-150 days.

5 Central (Malwa) highlands, Gujarat plains and Kathiawar peninsula, hot semi-arid ecoregion, with medium and deep black soils and GP 90-150 days

6 Deccan Plateau, hot semi-arid ecoregion, with mainly shallow and medium but also some deep black soils and GP 90-150 days.

7 Deccan Plateau of Telengana and Eastern Ghats, hot semi-arid ecoregion with red and black soils and GP 90-150 days.

8 Eastern Ghats, Tamil Nadu uplands and Deccan Plateau of southern Karnataka, hot semiarid ecoregion with red loamy soils and GP 90-150 days.

9 Northern plain, hot subhumid (dry) ecoregion, with alluvium-derived soils and GP 150-180 days.

10 Central highlands (Malwa, Bundelkhand and Eastern Satpura), hot subhumid ecoregion, with black and red soils and GP 150-180 days (up to 210 days in some places).

11 Eastern plateau (Chhatisgarh), hot subhumid ecoregion, with red and yellow soils and GP 150-180 days.

12 Eastern (Chhotanagpur) plateau and Eastern Ghats, hot subhumid ecoregion with red and lateritic soils, and GP 150-180 days (up to 210 days in some places).

13 Eastern Gangetic plain, hot subhumid (moist) ecoregion, with alluvium-derived soils and GP 180-210 days.

14* Western Himalayas, warm subhumid (to humid and perhumid) ecoregion, with alluviumderived soils and GP 210+ days.

15** Bengal and Assam Gangetic and Brahmaputra plains, hot subhumid (moist) to humid (and perhumid) ecoregion, with alluvium-derived soils and GP 210+ days.

16* Eastern Himalayas, warm perhumid ecoregion with brown and red hill soils and GP 210+ days

17* Northeastern hills (Purva chal), warm perhumid ecoregion with red and lateritic soils and GP 210+ days.

18 Eastern coastal plain, hot subhumid to semi-arid ecoregion, with coastal alluvium-derived soils and GP 90-210+ days.

19* Western ghats and coastal plain, hot humid-perhumid ecoregion with red, lateritic and alluvium-derived soils, and GP 210+ days.

20* Islands of Andaman-Nicobar and Lakshadweep hot humid to perhumid island ecoregion, with red loamy and sandy soils, and GP 210+ days.

* Indicates zones not included in the district level data.

** District level data contains Zone 13 districts in West Bengal but not Assam.

Source: NBSS&LUP, 1992

Table 2.3 Agroecological zones defined for this study

New zone	Description	Old zones	Number of districts covered
1	Northern Gangetic Plain (Punjab, Haryana, Uttar Pradesh, Bihar) with alluvial soils and growing season 90-180 days	9, 4 (Gangetic plains areas)	
2	Eastern Gangetic Plain (Uttar Pradesh, Bihar, West Bengal) with alluvial soils and growing season 150-210+ days	13, 15	
3	Central and Eastern Highlands (Madhya Pradesh, Orissa, Bihar and West Bengal) with black, red and lateritic soils; coastal areas of Orissa with alluvial soils; growing period 150-210 days	10, 11, 12, 18	
4	Central highlands, Gujarat plains and Deccan Plateau (Rajasthan, Gujarat, Madhya Pradesh, Maharashtra, Karnataka and Andhra Pradesh) with black soils; growing period 90-150 days.	5, 6, 4 (upland areas of Rajasthan and Gujarat)	
5	Deccan Plateau, Eastern Ghats and Tamil Nadu uplands (Andhra Pradesh, Karnataka, Tamil Nadu with mainly red soils; coastal areas of Andhra Pradesh and Tamil Nadu; growing period 90-150 days.	7, 8, 18	

Note: Desert areas, the Himalayas, the northeast, the Western Ghat and the west coast all drop out of the sample because they do not fall in our rainfall bounds. Zones omitted completely are 1, 14, 16, 17, 19. Zones 2 and 15 are omitted partially, and zones 3 and 18 are combined with neighboring zones.

What a New Typology Should Look Like

The existing agroclimatic typologies may be adequate for a narrow set of objectives, such as locating where certain crops are likely to be produced and which regions may be prone to certain natural resource management problems. Beyond such highly specific applications, however, the agroclimatic typologies are of limited use because they are so narrowly defined.

In order to be useful for designing a strategy to develop rainfed agriculture, a typology must be constructed on the basis of the whole range of factors that affect agricultural development. These extend far beyond simple agroclimatic conditions or even irrigation status. If a district has favorable growing conditions but lacks the infrastructure needed to support productive farming, for example, it should not be surprising to find poor performance in that district despite the favorable agroclimatic conditions. Later in the paper we will examine the determinants of agricultural performance and demonstrate that other factors in addition to agroclimatic conditions help explain the variation in performance. In addition to agroclimatic conditions and irrigation status, numerous additional variables can be hypothesized to influence agricultural development. Physical infrastructure such as roads and electrification, for example, and social infrastructure such as banks, markets and agricultural research and extension services, can be expected to play an important role in stimulating the agricultural sector. Demographic indicators such as population density and literacy levels also may be related to agricultural performance, as may economic policies that directly or indirectly affect input or output prices. Institutional considerations also may affect performance; they include laws governing trade, property rights, prices of inputs and outputs, etc., and the quality of services provided by government agencies.

Just as there are many determinants of performance of the agricultural sector, there also are many criteria for evaluating performance. Productivity growth is one that is commonly applied, but others include the levels of poverty and food security, the variability of production and income, and the degree of degradation of natural resources. The ideal typology of Indian agriculture would characterize regions or areas according to all the factors that determine performance over a broad range of criteria. In this way it could serve as a valuable planning tool for public investment in agricultural research, infrastructural development, poverty alleviation programs, policy and institutional reform, etc. While such a "super typology" might be unattainable, it presents an objective to work toward. In preparing this paper we lack the resources to develop an acceptable typology, but it remains a high priority for future research intended to support Indian rainfed agricultural development.

Later in the paper we analyze the determinants of rainfed agricultural development using multiple regression analysis. We will identify many of the agroclimatic and socioeconomic factors that contribute to performance of Indian rainfed agriculture according to a variety of criteria. We will stop short of creating a typology, but we will gain preliminary indications of the kinds of information that need to go into such a typology.

3

PERFORMANCE OF RAINFED AGRICULTURE: AN OVERVIEW

In this section we begin with some summary statistics of rainfed and irrigated agriculture, and then scrutinize differences in their growth rates for different crops over space and time.

Section 3 reviews the relevant literature, and section 4 presents an analysis based on district-level data.

IMPORTANCE OF RAINFED AGRICULTURE IN OVERALL SECTOR PERFORMANCE

Rainfed agriculture is clearly critical to agricultural performance in India. Nonetheless, it is difficult to precisely quantify the overall importance of the sector.

The widely quoted statistic is that 70% of cultivated area is rainfed, implying that rainfed agriculture is more important than irrigated agriculture.

However, this statistic grossly overstates the importance of rainfed agriculture in the economy for several reasons:

1. Since cropping intensity is lower in rainfed areas, the proportion of *gross cropped area* in rainfed areas was 66% in 1992.

2. Rainfed yields are on average less than half of irrigated yields (for food grains), so that the proportion of food grains produced in rainfed areas was 43% in the late 1980s (Planning Commission 1986).

   For non-food grains, the yield difference is even higher; we estimate that rainfed agriculture contributes less than half of the total value of production. Table 3.1 shows the area and production under rainfed and irrigated areas for various crops.

Table 3.1 Share of rainfed agriculture in area and production of major crops, India

Crops	Area (1987-89)	Production (1987-89)
Food grains	65.4*	NA
Rice	56.6	34.6
Wheat	22.7	11.0
Coarse Cereals	91.1*	NA
Jowar	95.2	78.7
Bajra	94.3	84.3
Maize	79.2	68.0
Pulses	90.2*	NA
Gram	79.9	62.1
Tur	95.4*	NA
Oilseeds	79.3*	NA
Groundnut	84.6	67.3
Rapeseed and Mustard	45.0	33.4
Cotton	69.1	7.8

Source: Agricultural Statistics at a Glance, and computed from Area and Production of PrincipalCrops in India

*Note: Data are for 1990-91

3. Farm size in rainfed areas is somewhat larger, so that the proportion of agricultural households that depend only on rainfed land is a little over half, considerably less than the proportion of cultivated area that is rainfed

4. Rainfed area as a proportion of total cultivated area is declining over time as land is converted to irrigated area. In 1956 about 17% of gross cropped area was irrigated compared to about 33% today.

Against this background, there may be counterbalancing factors that would increase the weight to rainfed areas in development strategies.

First, if sources of yield growth in irrigated areas are being exhausted and there are low returns to additional intensification in irrigated areas, then the potential role of rainfed areas in the future will increase.

In a later section, we briefly examine the evidence on yield potential in rainfed areas. Second, to the extent that other development objectives, especially poverty alleviation and conservation of the natural resource base, are important, rainfed areas merit increased attention relative to their weight in agricultural income generation.

In another section of this report, we have established that the poorest groups of the population depend on rainfed agriculture and that given the emphasis of the GOI and the Bank on poverty alleviation, rainfed agriculture deserves greater attention.

CROP YIELDS IN RAINFED AND IRRIGATED AGRICULTURE

A conventional wisdom that is widely held in the development community both inside and outside of India is that rainfed agriculture has been technologically stagnant.

In part, this arises from comparisons of yields and input use between rainfed and irrigated areas.

Dhawan, for example, has commented in several publications that the high yields in irrigated agriculture indicate the need for greater investment in the irrigated sector (Dhawan, 1988a).

Table 3.2, for example, shows Dhawan's comparison of irrigated and rainfed crop yields for the year 1983-84. In most of the country, irrigated yields surpass rainfed yields by about 1-2 tons/ha, though in the wetter states of the Himalayas, the northeast, and Kerala the difference is smaller.

These yield differences, while significant, are of only limited use because they do not control for differences in the composition of crops grown.

Table 3.3 presents average crop yields for irrigated and rainfed conditions in different states for the periods 1970-73, 1979-82, and 1986-89.

It shows that irrigated yields are generally much higher than rainfed yields, as expected. In many cases yield gaps are widening over time in favor of irrigated areas.

The difference is consistently high for cotton and quite high also for sorghum (though this excludes data for Maharashtra, which has the highest rainfed sorghum yield of any state). The difference is consistently smaller for gram.

For other crops, the gap varies substantially by state; more information would be needed on the agroclimatic conditions under which these crops are being grown in order to say more about the reasons for differences in relative yields. The yield comparisons between irrigated and rainfed agriculture are not surprising, but they should not be the sole basis for comparison between rainfed and irrigated agriculture.

Table 3.2 State-wise yields of irrigated and unirrigated segments, along with yield differential, 1983-84

State	Irrigated yield (kg/ha)			Unirrigated yield (kg/ha)			Yield differential (kg/ha)		
	Food grains	Non-food grains	Total	Food grains	Non-food grains	Total	Food grains	Non-food grains	Total
Andhra Pradesh	2083	3697	2257	684	1056	796	1399	2621	1461
Tamil Nadu	1938	5002	2373	730	1107	852	1208	3895	1521
UP	1914	4628	2292	983	772	977	931	3856	1315
MP	1566	2745	1637	945	907	943	621	1838	694
Punjab	2999	3741	3025	1359	816	1314	1640	2925	1711
Haryana	2258	4070	2328	583	1068	606	1675	3002	1722
Gujarat	2291	2517	2364	930	919	926	1361	1598	1438
Rajasthan	1519	1466	1509	635	1143	661	884	323	848
Maharashtra	1285	6254	2563	752	1031	767	533	5223	1796
Kamataka	2377	5622	3058	803	840	808	1574	4782	2250
Bihar	1412	2562	1429	838	2310	890	574	252	539
Orissa	1651	3519	1755	964	901	961	687	1618	794
West Bengal	1953	1295	1918	1092	617	1078	861	678	840

Kerala	1795	NA	1795	1521	NA	1521	274	NA	274
Assam	1194	NA	1194	1052	NA	1052	142	NA	142
HP	1588	NA	1588	1480	NA	1480	108	NA	108
J & K	1917	NA	1917	1012	NA	1012	905	NA	905
Average	1980	3979	2208	864	1001	877	1116	2978	1331

Source: Dhawan (1988)

Note: Total yield, as also non-ffod grain yield, is in food energy equivalents (FEES).

Table 3.3 State-wise irrigated and rainfed yields of major crops and the share of rainfed agriculture in area and production

Crop	1970-73			1979-82			1986-89			1986-89 Share of rainfed Ag.Irrigated Yield	
	Irrigated Yield	Rainfed Yield	Irrigated Yield/ Rainfed Yield (%)	Irrigated Yield	Rainfed Yield	Irrigated Yield/ Rainfed Yield (%)	Irrigated Yield	Rainfed Yield	Irrigated Yield/ Rainfed Yield (%)	Area in %	Prdn. in %
Rice											
Andhra Pradesh	1512	691	219	1986	833	238	2134	746	286	5.8	2.9
Assam				1353	805		1412	778	181	66.2	54.7
Bihar	880	533	165	1278	621	206	1258	791	159	64.2	56.0
Gujarat	1762	1144	154.	1772	947	187	2232	983	227	57.0	35.1
Himachal Pradesh	1337	945	141	1454	956	152	1263	1229	103	45.8	45.9

Karnataka	1771	1243	142	1852	1544	120	2350	1336	176	39.0	23.6
Kerala	1577	1242	127	1430	1290	111	1775	1433	124	55.0	27.1
Madya Pradesh	1095	741	148	1083	633	171	1454	793	183	79.0	71.4
Maharashtra	909	767	119	1535	1443	106	1305	1207	108	74.2	68.0
Orissa	1007	910	111	1153	653	176	1139	688	166	66.8	43.0
Punjab	1979	1296	153	2460	1018	242	3278	1155	284	1.2	0.4
Rajasthan				1511	553	273	2204	1129	195	58.5	57.7
Tamil Nadu	1990	1012	197	2064	862	239	3256	969	336	8.5	1.7
Uttar Pradesh				1247	758	164	1581	1163	136	66.8	54.9
West Bengal	1683	1012	166	1513	784	193	1663	1013	164	75.4	46.9
Jowar											
Andhra Pradesh	523	292	179	1316	573	230	1511	589	257	98.4	92.9
Gujarat	829	221	375	1302	467	279	680	278	245	97.2	78.6
Haryana	265	260	102	316	214	147	318	109	292	67.1	27.7
Karnataka	1329	720	185	1132	876	129	2677	1100	243	93.0	60.9
Madya Pradesh		187		1042	703	148		758		99.9	98.6
Maharashtra		419			1050			1105		94.2	69.7
Tamil Nadu	1313	584	225	1638	748	219	2425	908	267	92.7	93.9
Bajra											
Andhra Pradesh	1050	367	286	1508	586	257	1031	508	203	89.0	77.8
Gujarat	1100	717	153	1302	878	148	1026	572	179	90.9	79.7

Haryana	843	594	142	845	557	152	917	531	173	84.6	96.7
Karnataka	949	174	545	689	361	191	1064	482	221	91.5	95.1
Maize											
Andhra Pradesh	1560	856	182	2554	1546	165	2493	1155	216	79.5	63.2
Bihar	1013	554	183				2406	1160	207	70.3	52.7
Gujarat	1103	1016	109	1911	913	209	1557	1007	155	93.5	99.3
Haryana	1041	1031	101	883	815	108	1050	949	111	68.5	48.4
Himachal Pradesh	2339	1752	134	2034	1822	112	2439	1260	194	92.7	54.8
Karnataka				2843	2923	97	3010	1780	169	20.0	15.4
Madya Pradesh	2181	929	235	1518	563	269	1222	1106	110	98.7	74.8
Maharashtra				1108	983	113	1984	832	238	48.1	33.3
Punjab	1632	1346	121	1755	1232	142	1692	1373	123	41.4	31.6
Barley											
Bihar	722	581	124	885	648	137				87.5	87.2
Haryana	1437	913	157	1526	860	178		927		27.5	14.2
Himachal Pradesh	1734	1268	137	1196	1214	98	1844	1044	177	84.8	49.0
Madya Pradesh	1406	858	164	1378	781	176	1199	842	142	77.9	85.0
Punjab	1255	787	159	1800	1060	170	1283	1031	124	15.4	7.2
Rajasthan	1145	778	147	1373	688	199	2356	1165	202		
Uttar Pradesh	1139	922	124	1309	904	145	1661	927	179	51.6	27.5
West Bengal				812	806	101					

Wheat											
Assam				1346	1154	117	1027	1092	94	100.0	87.4
Bihar	1011	721	140	1313	932	141	1510	1554	97	21.0	18.7
Gujarat	1751	583	301	2327	492	473	2325	429	542	24.3	4.9
Haryana	2037	1287	158	2366	1479	160	3054	1466	208	3.7	1.9
Himachal Pradesh	1369	1022	134	1636	1170	140					
J & K	1465	657	223	1246	951	131					
Karnataka	653	220	297	1209	395	306	1162	335	347	72.0	55.8
Madya Pradesh	1326	660	201	1405	707	199	2021	816	248	59.8	41.3
Maharashtra				1107	524	211	1233	599	206	44.1	20.5
Punjab	2393	1112	215	2887	1489	194	3310	1752	189	5.1	3.3
Rajasthan	1441	749	192	1622	808	201	2159	1257	172	9.8	6.4
Uttar Pradesh	1422	910	156	1667	948	176	2006	1257	160	14.1	9.1
West Bengal				1429	971	147					
Gram											
Gujarat	1127	666	169	1027	635	162	877	489	179	73.6	65.2
Haryana	845	553	153	552	450	123	733	527	139	66.1	41.3
Karnataka	348	123	283	588	438	134	512	429	119	91.7	11.2
Madya Pradesh	812	623	130	931	536	174	825	640	129	82.9	80.5
Maharashtra				530	332	160	567	383	148	77.0	67.1

Punjab	870	722	120	622	477	130	623	477	131	86.8	54.9
Rajasthan	887	564	157	814	522	156	806	716	113	74.9	75.2
Uttar Pradesh	823	750	110	760	637	119	1041	716	145	82.6	75.9
Groundnut											
Andhra Pradesh	1016	719	141	1160	705	164	948	808	117	81.5	69.8
Gujarat	1051	927	113	929	791	117	1593	680	234	91.6	68.2
Karnataka	1616	1345	120	964	629	153	970	647	150	79.7	65.3
Madya Pradesh				630	642	98	1189	923	129	94.3	91.2
Maharashtra				1400	682	205	967	752	129	97.3	78.3
Punjab	980	1022	96	1151	933	123	920	412	223	47.5	22.1
Rajasthan	1388	592	235	951	510	187	1195	761	157	67.8	64.9
Tamil Nadu	1539	903	170	1649	820	201	1838	943	195	73.8	61.4
Cotton											
Andhra Pradesh	308	70	440	387	184	211	746	311	240	77.8	16.3
Gujarat	306	165	185	358	133	269	387	89	435	67.1	3.9
Karnataka	134	23	583	310	83	374	419	122	343	81.6	4.8
Madya Pradesh	140	73	192				293	110	266	11.0	14.1
Maharashtra	246	46	535	210	86	245	293	76	386	96.0	9.0
Rajasthan	217	106	205	213	63	338	391	206	190	5.8	0.5
Tamil Nadu	312	71	439	375	77	485	445	167	266	59.7	5.7

Source: Area and Production of Principal Crops in India, Various issues

Rainfed yields and input use will always lag behind those in irrigated areas, so other *performance measures must be used.*

In fact, there are many indicators of remarkable success in technology adoption and yield growth in rainfed agriculture in India. Examples include the following:

- The rapid shifts of cropping patterns and adoption of new crops in rainfed areas (Kelley and Parthasarathy Rao 1994). The recent widespread adoption of new oilseed crops, especially soyabeans and sunflowers, in central and southern India is testimony to the potential for rapid change in rainfed areas, given appropriate technology, input and marketing support, and policy incentives (Singh et al). The rapid and broadly based growth of the cotton sector is further testimony to the potential for these types of changes.
- The relatively high growth performance in yields of some crops that are grown largely under rainfed conditions. This is most apparent for cash crops, especially cotton and oilseeds but is also evident for some food grains in some states (e.g. sorghum in Maharashtra, pearl millet in Gujarat, finger millet (ragi) in Karnataka, and maize and pigeon pea in some districts).
- The overall growth of total factor productivity of 1% annually in agroclimatic zones that depend largely on rainfed agriculture (Evenson, Pray and Rosegrant, 1995). Although this growth rate lags behind the 1.5% rate observed in the Green Revolution areas of northwest India, the difference is surprisingly small.

Nonetheless, the overall growth of rainfed agriculture has been slow. This is evident in overall statistics for coarse grains and especially pulses, and rising real prices for some crops such as pulses that are largely grown in rainfed areas (Kelley and Parthasarathy Rao, 1994).

In a few states where there are fairly complete and reliable data on irrigated and rainfed yields, the growth rate of rainfed yields has generally lagged. For example, for the period 1960-85, rainfed wheat yields increased at a rate of 1.4% annually, only half of that observed in irrigated areas (Byerlee, 1992).

Yields of rainfed rice lagged irrigated yields until 1980 but showed remarkable growth thereafter, especially in West Bengal (table 3.3 and table 3.4). Table 3.4 shows changes in growth rates over time under irrigated and rainfed conditions between 1970 and 1989. Patterns vary by crop and region, and clear patterns are difficult to discern.

ADOPTION OF MODERN INPUTS: HYVS AND FERTILIZER

The significant success stories for rainfed areas reflect widespread adoption of modern inputs, especially improved seed and fertilizers. One of the best kept secrets of Indian agriculture (at least from the point of view of an outside observer of Indian agriculture) has been the remarkable spread of HYVs in rainfed areas beginning in the mid-1970s and accelerating in the past decade.

In 1976, 84% of the HYV area was under irrigation (calculated from table 27 of Desai, 1982). By the early 1990s over 40% of the area of HYVs was sown in rainfed areas. From about 1976, the cereal area sown to HYVs has exceeded the irrigated area of cereals, and this gap has widened over time.

We estimate that in the 1980s, an additional 22 million ha of cereal area was sown to HYVs. Of this amount, 16 million ha or nearly three quarters of the total area expansion occurred in rainfed areas.

This represents the most spectacular example of widespread adoption of HYVs by small-scale farmers under rainfed conditions in the world. The following statistics and assumptions, drawn mainly from table 3.5 (NCAER 1990), elaborate on these calculations.

- The largest expansion in area of HYVs has been for rice in eastern India. HYVs are now sown on 50% of the rice area in Eastern India compared to only 20% of the rice area that is irrigated.
- The next largest expansion has been in coarse grains, nearly all under rainfed conditions, especially sorghum in Maharashtra, pearl millet in Gujarat and ragi (finger millet) in Karnataka. Much of this area expansion has been through adoption of hybrids produced by both the public and private sectors (Pray et al, 1991).
- Although use of HYVs is undoubtedly greater in better rainfall zones, there have been notable successes in some dry areas, especially the cases of millet and ragi (finger millet) noted above.
- Similarly in cash crops, adoption of improved varieties of cotton has been widespread (an estimated 86% of the rainfed cotton area), including the sowing of hybrid cotton on nearly 3 million ha (NCAER, 1990; Basu, Narayanan and Singh, 1992). The expansion of nontraditional oilseeds has also taken place using newly introduced varieties and hybrids, and about half of the groundnut area is now sown to HYVs.
- The most rapid gains in yields of rainfed crops have usually been associated with areas of expansion of HYVs.

Table 3.4 Growth rates of state-wise irrigated and rainfed crop yields

	Growth Rates (%)					
	1970-73 - 1979-82		1979-82 - 1986-89		1970-73 - 1986-89	
Crop	Irrigated Yield	Rainfed Yield	Irrigated Yield	Rainfed Yield	Irrigated Yield	Rainfed Yield
Rice						
Andhra Pradesh	3.08	2.11	1.03	-1.56	2.18	0.48
Assam			0.62	-0.48		
Bihar	4.23	1.71	-0.23	3.53	2.26	2.50
Gujarat	0.07	-2.08	3.35	0.53	1.49	-0.95
Himachal Pradesh	0.94	0.12	-1.99	3.65	-0.35	1.65
Karnataka	0.50	2.43	3.46	-2.04	1.78	0.45
Kerala	-1.08	0.42	3.14	1.52	0.74	0.90
Madya Pradesh	-0.12	-1.74	4.30	3.28	1.79	0.43
Maharashtra	5.99	7.29	-2.29	-2.52	2.29	2.88
Orissa	1.51	-3.62	-0.17	0.75	0.77	-1.73
Punjab	2.45	-2.64	4.18	1.82	3.20	-0.72
Rajasthan			5.54	10.73		
Tamil Nadu	0.40	-1.77	6.73	1.69	3.12	-0.27
Uttar Pradesh			3.44	6.30		
West Bengal	-1.18	-2.80	1.36	3.74	-0.07	0.01

Jowar						
Andra Pradesh	10.80	7.77	1.99	0.40	6.86	4.48
Gujarat	5.14	8.67	-6.96	5.51	-1.23	7.28
Haryana	1.96	-2.11	0.11	-9.21	1.15	-5.28
Karnataka	-1.77	2.20	13.08	3.31	4.47	2.69
Madya Pradesh		15.86		1.08		9.14
Maharashtra		10.77		0.73		6.26
Tamil Nadu	2.49	2.78	5.77	2.81	3.91	2.80
Bajra						
Andhra Pradesh	4.10	5.33	-5.29	-2.01	-0.11	2.05
Gujarat	1.88	2.28	-3.34	-5.94	-0.44	-1.40
Haryana	0.03	-0.71	1.18	-0.69	0.53	-0.70
Karnataka	-3.49	8.44	6.40	4.23	0.72	6.58
Madya Pradesh	13.68	-4.64		7.82		0.62
Maharashtra	8.07	10.57	0.06	-2.04	4.49	4.87
Punjab	0.06	-0.70	2.00	-0.04	0.90	-0.41
Rajasthan			-0.05	13.60		
Tamil Nadu	3.55	3.65	3.37	5.60	3.47	4.50
Maize						
Andhra Pradesh	5.63	6.79	-0.34	-4.08	2.97	1.89

Bihar					5.55	4.73
Gujarat	6.30	-1.18	-2.89	1.41	2.18	-0.06
Haryana	-1.82	-2.57	2.51	2.19	0.05	-0.51
Himachal Pradesh	-1.54	0.44	2.63	-5.13	0.26	-2.04
Karnataka			0.82	-6.84		
Madya Pradesh	-3.95	-5.41	-3.05	10.12	-3.56	1.10
Maharashtra			8.68	-2.35		
Punjab	0.81	-0.97	-0.52	1.56	0.22	0.13
Rajasthan	-2.65	-3.77	0.34	9.43	-1.36	1.79
Barley						
Bihar	2.29	1.23				
Haryana	0.67	-0.66				
Himachal Pradesh	-4.05	-0.48	6.38	-2.14	0.39	-1.21
Madya Pradesh	-0.22	-1.05	-1.96	1.09	-0.99	-0.12
Punjab	4.09	3.36	-4.72	-0.40	0.14	1.70
Rajasthan	2.04	-1.34	8.02	7.81	4.61	2.56
Uttar Pradesh	1.55	-0.22	3.46	0.36	2.38	0.03
West Bengal						
Wheat						
Assam			-3.79	-0.79		

Bihar	2.95	2.88	2.02	7.59	2.54	4.91
Gujarat	3.21	-1.86	-0.01	-1.94	1.79	-1.90
Haryana	1.67	1.56	3.72	-0.13	2.56	0.82
Himachal Pradesh						
Jammu & Kashmir						
Karnataka	7.08	6.73	-0.56	-2.31	3.67	2.68
Madya Pradesh	0.65	0.77	5.33	2.06	2.67	1.33
Maharashtra			1.55	1.94		
Punjab	2.11	3.30	1.97	2.35	2.05	2.88
Rajasthan	1.32	0.85	4.17	6.52	2.56	3.29
Uttar Pradesh	1.78	0.46	2.68	4.11	2.17	2.04
West Bengal						
Gram						
Gujarat	-1.03	-0.52	-2.23	-3.67	-1.56	-1.91
Haryana	-4.62	-2.26	4.13	2.27	-0.89	-0.30
Karnataka	6.00	15.16	-1.96	-0.30	2.44	8.12
Madya Pradesh	1.53	-1.64	-1.71	2.56	0.10	0.17
Maharashtra			0.96	2.05		
Punjab	-3.66	-4.51	0.03	0.00	-2.06	-2.56
Rajasthan	-0.94	-0.86	-0.15	4.63	-0.59	1.51

Uttar Pradesh	-0.89	-1.79	4.60	1.68	1.48	-0.29
West Bengal						
Groundnut						
Andhra Pradesh	1.48	-0.21	-2.84	1.96	-0.43	0.73
Gujarat	-1.36	-1.75	8.01	-2.14	2.63	-1.92
Karnataka	-5.58	-8.09	0.09	0.40	-3.14	-4.47
Madya Pradesh			9.50	5.33		
Maharashtra			-5.15	1.42		
Punjab	1.80	-1.01	-3.15	-11.02	-0.39	-5.52
Rajasthan	-4.11	-1.64	3.31	5.88	-0.93	1.59
Tamil Nadu	0.77	-1.07	1.57	2.02	1.12	0.27
Cotton						
Andhra Pradesh	2.56	11.31	9.83	7.81	5.68	9.77
Gujarat	1.75	-2.37	1.13	-5.58	1.48	-3.78
Karnataka	9.78	15.33	4.38	5.66	7.39	10.99
Madya Pradesh	4.72	2.60				
Maharashtra	-1.73	7.20	4.85	-1.75	1.10	3.19
Rajasthan	-0.22	-5.62	9.09	18.44	3.75	4.24
Tamil Nadu	2.07	0.95	2.46	11.63	2.24	5.49

Source: Computed from Table 3.

Nonetheless, about half of the rainfed area is still sown to traditional varieties. The adoption of HYVs of pulses has been minimal, and for post-rainy season crops, especially rabi sorghum and wheat in Central and Southern India, the use of HYVs is negligible. We will return below to the special problems of developing HYVs for difficult rainfed areas. Fertilizer use in rainfed areas has expanded rapidly, especially since 1976. In that year, less than 20% of the rainfed area was fertilized and the average dose per ha of rainfed area was only 9 kg/ha (Desai, 1982). By 1989, over half of the area was fertilized with an average application of 34 kg/ha of rainfed area (table 3.5).

The use of fertilizer on rainfed crops is generally associated with use of HYVs. Again, the lowest use of fertilizer is in pulses and the highest use is on cash crops, such as cotton. Other modern inputs have also been adopted fairly widely for some crops, especially pesticide use on cotton and hybrid sorghum. However, no comprehensive data set exists on use of these inputs. To summarize this section, patterns in rainfed agriculture clearly show evidence that it is not a technologically stagnant sector, though regional variations exist.

In the next section we examine growth rates in different categories of rainfed agriculture in order to see how changes in HYV and fertilizer adoption have led to changes in growth rates of output and other performance indicators, and to get a better understanding of the circumstances under which rainfed agriculture has performed relatively well or poorly.

Ahluwalia 2 also adjusts his analysis to control for the effects of changes in rainfall patterns and finds that increasing growth in the 1980s and 1990s was not the result of changes in rainfall. 3Studies of TFP include Desai (1994), Dholakia and Dholakia (1993), Rosegrant and Evenson (1994), Kumar and Rosegrant (1994), Kumar and Mruthyunjaya (1992), and Sidhu and Byerlee (1991).

Table 3.5 Irrigation, HYV and fertilizer status of major crops, all India, 1989

	Percent Irrigated	Percent HYV	Percent area HYV		Percent fertilized		Fertilizer use per ha fertilized (kg/ha)			Average fertilizer (kg/ha)	
Rainfed			Irrigated	Rainfed	Irrigated	Rainfed	All	Irrigated	Rainfed	Irrigated	
Rice - K	57	68	88	42	94	57	78	116	57	109	32
Wheat	86	78	85	34	99	62	94	135	54	133	33
Sorghum	17	56	72	52	83	63	67	67	57	55	36
Pearl millet	20	57	80	52	81	48	55	60	43	49	21
Maize	49	55	73	38	95	74	84	83	52	79	38
Finger millet	13	33	67	28	89	71	73	124	72	110	51
Pulses-K	8	12	31	10	74	39	42	64	53	47	20
Pulses-R	38	14	19	10	75	37	52	66	41	50	15
G/nuts-K	32	51	76	40	76	82	80	86	61	65	50
G/nuts-R	62	79	83	57	89	74	83	114	85	101	63
Oilseeds	22	39	54	35	94	72	76	63	45	59	32
Cotton	47	91	98	86	98	75	85	123	79	120	59
Sugarcane	99	86	92	72	98	68	98	168	129	165	88
Tobacco	72	60	45	56	99	94	98	165	119	164	111
All Kharif	n.a.	n.a.	n.a.	n.a.	86	56	73	88	33	76	19
All Rabi	n.a.	n.a.	n.a.	n.a.	87	51	79	111	35	96	18

Source: NCAER 1990

PATTERNS OF OUTPUT AND YIELD GROWTH BEFORE AND SINCE THE GREEN REVOLUTION

Several studies have examined agricultural production patterns before and since the green revolution of the mid-1960s. The key findings of these studies can be summarized as follows:

- Growth rates in production were not significantly different before and after the green revolution. Output growth was triggerd by increased cropped area in the pre-green revolution era (before the mid-1960s) but increases in yield thereafter. Studies by Vaidyanathan (1993), Hanumantha Rao (1994), Ramakrishna (1993) and Ahluwalia (1995) all agree on this point. While different studies all show slightly different numbers depending on the data and methods used, table 3.6 gives an idea of the figures, using 1965 as the cutoff point before the green revolution.
- Within the period after the green revolution, growth in agricultural output was significantly higher after about 1980 than before (Hanumantha Rao, Ahluwalia.) Table 3.7 shows Ahluwalia's output growth rate calculations based on state-level data, and table 3.8 shows his estimates of yield growth rates.2 Table 3.9 shows growth rates of cropped area.
- The higher growth in output and yields after 1980 results from at least two factors. First, HYVs continued to spread after 1980, particularly in rice and coarse grains for which the use of HYVs first spread slowly (table 3.10). Second, input use increased significantly in the 1980s, partly due to subsidies for inputs such as power, water and fertilizer (Ahluwalia, 1995; Repetto, 1993). Ahluwalia summarized several recent studies that show a decline in the growth of total factor productivity (TFP).3 This finding may signal that acceleration in growth rates may have taken place at the expense of the natural resource base.

- Breaking down output growth rates by crops, Hanumantha Rao (1994) showed that rice performed particularly well after about 1980. Ahluwalia (1995) found high growth rates in wheat and rice and non-foodgrains but very low growth in coarse grains and pulses.
- Breaking down growth rates by regions, Hanumantha Rao (1994) found the highest growth rates in the period 1978-79 to 1988-89 in the traditional green revolution states of Punjab and Haryana, but also in rice-growing areas of eastern India. Ahluwalia's (1995) analysis found similar results for the period 1982-1994, with some differences resulting from the different data set and years. Ahluwalia's estimated growth rates of the value of agricultural output by state are shown in table 3.11.
- Ahluwalia's summary of studies of TFP suggest that the main sources of TFP growth are agricultural research, education, extension, market infrastructure, irrigation and mechanization.

Table 3.6 Annual growth in food grain production, area and productivity 1950-51 to 1990-91

Period	Food grains			Non-food grains			All crops		
	Production	Area	Yield	Production	Area	Yield	Production	Area	Yield
1950-51 to 1964-65	2.58	1.24	1.32	3.41	2.18	1.20	2.75	1.43	1.31
1967-68 to 1990-91	2.80	0.20	2.60	2.82	0.54	2.27	02.82	0.27	2.53
1950-51 to 1990-91	2.71	0.55	2.16	2.76	0.91	1.75	2.69	0.63	2.05

Source: Ramakrishna, 1993.

Note: Ramakrishna omits the severe drought years of 1965-66 and 1966-67 because as starting or ending points in the analysis, they would move growth rates excessively downward or upward.

Table 3.7 Trend growth rates of production of major crop groups

	Weight in Index of Agricultural Production (Base 1981-82)	Entire Period 1951-94	Pre-Green Revolution 1951-67	Post-Green Revolution 1968-94	Significant Change Between Pre- and Post-Green Revolution at 5% Level	Percent Contribution to All Crop Growth in Post-Green Revolution
			(percent per annum)			
All Crops Index	100.00	2.6	2.4	2.8	No	100.0
Foodgrains	62.92	2.5	2.0	2.6	No	58.4
Total Cereals	54.98	2.9	2.5	2.9	No	56.9
Course Cereals	10.79	1.1	1.7	0.7	Yes	2.7
Pulses	7.94	0.5	0.2	0.9	No	2.6
Non-Foodgrains	37.08	2.8	3.2	3.1	Yes	41.1
Oilseeds	12.64	2.5	2.5	3.3	Yes	14.9
Fibers	5.09	2.2	3.3	2.4	No	4.4
Sugarcane	8.11	3.0	4.0	2.9	No	8.4
Plantation Crops	2.29	3.5	2.9	3.5	Yes	2.9
Condiments & Spice	2.59	2.1	0.8	3.2	Yes	3.0
Fruits and Vegetables	4.90	5.1	7.6	3.9	Yes	6.8
GDP Agriculture (including forestry and fishing)		2.4	2.0	2.6	No	

Source: Data from GOI, Ministry of Agriculture; CSO

Table 3.8. Trend growth rates of yield of major crop groups

	Weight in Index of Agricultural Production (Base 1981-82)	Post-Green Revolution 1968-94	Post-Green Revolution Period-I 1968-81	Post-Green Revolution Period-II 1982-94	Significant Change Between Period I and Period II at 5% Level
		(percent per annum)			
All Crops Index	100.00	2.0	1.3	2.5	Yes
Foodgrains	62.92	2.2	1.3	2.8	Yes
Total Cereals	54.98	2.4	1.7	2.9	Yes
Coarse Cereals	10.79	1.7	1.6	2.4	No
Pulses	7.94	0.7	-0.7	1.1	Yes
Non-Foodgrains	37.08	1.8	1.2	2.3	Yes
Oilseeds	12.64	1.6	0.5	2.3	Yes
Fibers	5.09	2.6	2.3	4.0	No
Sugarcane	8.11	1.3	0.8	1.5	No
Plantation Crops	2.29	1.8	2.3	2.2	No
Condiments & Spice	2.59	1.5	0.4	1.8	Yes
Fruits & Vegetables	4.90	1.9	2.0	2.0	No

Source: Data till 1991 from GOI, Ministry of Agriculture, 'Area and Production of Principal Crops in India 1990-93.' Data from 1992-94 from Ministry of Agriculture

Table 3.9. Trend growth rates of area of major crop groups

	Weight in Index of Agricultural Production (Base 1981-82)	Post-Green Revolution 1968-94	Post-Green Revolution Period-I 1968-81	Post-Green Revolution Period-II 1982-94	Significant Change Between Period I and Period II at 5% Level
		(percent per annum)			
All Crops Index	100.00	0.4	0.5	0.2	No
Foodgrains	62.92	0.1	0.4	-0.4	Yes
Total Cereals	54.98	0.1	0.4	-0.4	Yes
Coarse Cereals	10.79	-1.2	-1.0	-1.9	Yes
Pulses	7.94	0.2	0.4	-0.3	No
Non-Foodgrains	37.08	1.3	0.9	1.9	Yes
Oilseeds	12.64	1.2	0.3	2.6	Yes
Fibers	5.09	-0.3	0.2	-0.6	No
Sugarcane	8.11	1.6	1.8	1.4	No
Plantation Crops	2.29	2.2	2.5	2.2	Yes
Condiments & Spice	2.59	1.6	1.6	1.3	No
Fruits & Vegetables	4.90	1.7	2.3	1.4	Yes

Source: Data until 1991 from GOI, Ministry of Agriculture, 'Area and Production of Principal Crops in India 1990-93.' Data from 1992-94 from Ministry of Agriculture.

Table 3.11 Trend growth rates of state domestic product from agriculture

Region/States	*Share in All India Net Domestic Product from Agriculture (1989-90)*	*Share of Agriculture in State Domestic Product (1989-90)*	*Growth Post-Green Revolution 1968-93*	*Growth Post-Green Revolution Period I 1968-81*	*Growth Post-Green Revolution Period II 1982-93*	*Significant Change Between Period I and Period II at 5% Level*
			(percent per annum)			
North	10.2	44.3	3.8	3.1	4.8	Yes
Punjab	6.3	45.2	4.0	3.3	4.8	Yes
Haryana	3.9	42.9	3.5	2.9	4.8	No
Uttar Pradesh	15.4	40.7	2.8	1.9	2.9	No
East	20.6	35.2	2.6	1.9	3.8	Yes
Assam	2.6	36.6	1.9	1.2	1.7	No
Bihar	7.1	39.8	1.5	1.5	1.2	No
Orissa	3.8	45.0	3.1	2.3	3.4	No
West Bengal	7.1	28.3	3.8	3.0	5.3	Yes

Center	12.6	38.9	2.9	1.5	3.6	No
Mayda Pradesh	7.1	36.7	2.5	0.3	2.9	Yes
Rajasthan	5.4	42.3	3.4	3.4	4.4	No
West	14.4	22.5	2.6	3.8	1.6	No
Gujarat	5.2	27.0	1.8	3.2	-1.3	No
Maharashtra	9.2	20.6	3.0	4.2	3.3	No
South	21.2	29.3	2.0	1.7	2.7	No
AP	8.3	35.2	2.5	2.2	1.6	No
Karnataka	5.6	33.2	2.6	2.7	2.7	No
Kerala	2.8	28.9	1.3	0.4	4.4	Yes
Tamil Nadu	4.5	20.4	1.2	0.9	3.6	Yes
All India	100.0	29.8	2.7	2.1	3.1	No

Notes: (i) To conform with the post-1980/81 data from the new CSO series, data from 1968 to 1981 are converted to an 1980-81 base from the old CSO series with 70-71 base. (ii) For Orissa data were available only until 1990. The shares, in the first two columns, for this state are averages of 1988-90; the growth rates are for 1968-90, 1968-81 and 1982-90 respectively. By corollary, these periods apply to the eastern region too. (iii) The All India figures are for NDP from agriculture.

Source: Data from C.S.O.

Table 3.10 Spread of high-yield varieties (HYVs): all India

Year	Paddy	Wheat	Sorghum	Pearl Millet	Maize	Total
1966-67	2.5	4.2	1.1	0.5	4.1	2.3
1970-71	14.9	35.5	4.6	1.6	7.9	16.7
1975-76	31.5	65.8	12.2	25.0	18.8	34.1
1980-81	45.4	72.3	22.1	39.2	26.7	44.9
1985-86	57.1	83.0	37.8	46.8	31.0	55.4
1987-88	58.1	85.4	38.7	45.4	38.8	54.1
1992-93	65.8	88.2	53.1	53.0	43.2	67.0

Source: Parikh, Mahendra Dev and Deshpande (1993) (calculated from Fertilizer Statistics, and Area and Production of Principal Crops, Ministry of Agriculture).

4

GROWTH OF OUTPUT, YIELDS AND CROPPED AREA: A DISTRICTLEVEL ANALYSIS

In this section we compare growth rates of value of output in irrigated and unirrigated areas using district level data. The analysis in this section proceeds in two steps. First, we conduct tabular analysis of the growth of value of output for irrigated and rainfed districts of the five agroclimatic zones listed in table 2.3.

This analysis is based on district level data for 243 districts, the majority of the districts in the five zones. Figures are presented for three different time periods to examine differences in summary performance indicators over space and time. Second, we estimate a production function based on the district level data. The production function follows ICAR's agroecological zoning system; of the 20 zones defined by ICAR, 15 are included in the area under study.

Based on the background information provided in the previous section, we expect to find in the analysis the following trends. In the production function analysis, we expect irrigation to make the greatest contribution to the value of output. In the tabular analysis in which the years before, during and after the green revolution are examined separately, we expect the trends to change as follows. In period 1, prior to the green revolution, we expect rising growth rates to be driven by increases in both net and gross cropped area. Net cropped area rises as more land is cleared for agriculture, and gross cropped area increases with the spread of irrigation and multiple-cropping in dryland areas. In addition, irrigation will raise crop yields somewhat even without the benefit of HYVs. In period 2, we expect growth rates to be triggered by large yield increases in irrigated areas. In period 3, we expect growth rates in irrigated areas to slow but growth rates in favorable rainfed areas to rise. In periods 2 and 3 we do not expect much overall change in net cropped area, though of course we do expect shifts in area from one crop to another.

Defining three subperiods in the study period of 1956-1991 necessarily is somewhat arbitrary. Ideally the divisions would be based on changes in the factors expected to affect growth in value of output, such as the spread of HYVs or irrigation. On the other hand, easily observable changes in such variables may not exist. Other factors will also be important when we calculate growth rates for each of the subperiods. For example, the calculated growth rate may rise significantly if the first year of a subperiod is a drought year, and it may fall if the last year is a drought year. As a result, working around drought years is more important than identifying changes in trends in growth rates and HYV adoption. The three subperiods defined in the study are 1956/57-1967/68, 1968/69-1979/80, and 1980/81-1990/1991. The first period extends until 1968/69 even though 1965/66, the year HYVs were introduced, would be a natural starting point for period 2. However, 1965/66 and 1966/67 were both severe drought years. If the last year of a period is a drought year with unusually low output, the growth rate for the period as a whole will be underestimated. Likewise, if the first year is a drought year, the growth rate for the whole period will be overestimated. At the end of the first period, HYVs covered roughly 5% of all area under cereals. At the end of the second period, HYVs covered nearly half of all cereal area, and about three quarters of all wheat area. We expect that almost all of the growth in HYVs in period 2 took place in irrigated areas, and much of that in period 3 took place in rainfed areas. While the definition of subperiods may not perfectly capture changes in the status of HYVs, it is acceptable.

An alternate approach would be to delete the two drought years, ending period 1 in 1964/65 and beginning period 2 in 1967/68. This is the same approach as used by Hazell (1982), Hanumantha Rao (1994) and Ramakrishna (1993) in similar studies, cited above. In fact, a number of reasonable divisions could have been devised, but none would be perfect. It is important to note that every delineation of time periods will yield slightly different results. The results reported below represent rough indications of the growth rates in output, yield and cropped area for different regions and conditions and time periods in the overall period under study, but they are not the last word. More detailed analysis will be needed for more precise results. 4Net cropped area and cropping intensity are the appropriate data for the analysis of all crops. For cereals and other crops, gross cropped area is sufficient and cropping intensity may be omitted. "Output" refers to the value of output for the analysis of all crops and all cereals, but physical output is used for the analysis of individual crops.

TABULAR ANALYSIS OF GROWTH RATES IN VALUE OF OUTPUT

For the tabular analysis of the sources of growth of output, we group districts into a total of ten categories, including five agroclimatic zones and two irrigation categories as described in table 2.3. We conduct the analysis separately for all crops, all cereals, and major cereal crops. The tables contain growth rates of value of output, yield, net cropped area and cropping intensity.

Note that because production$^{(a)}$ = net area$^{(A)}$ x yield$^{(Y)}$ x cropping intensity$^{(C)}$,

then

growth (a) = growth (A) + growth (Y) + growth (C).

The figures in the tables reflect this disaggregation, as the growth of output is always the sum of the growth rates of the other three factors. As explained in section 2, irrigated area grew steadily during the period under study, so some districts changed from less than 25% area irrigated to greater than 25% area irrigated. This is the case for 59 districts out of the 243 districts in the sample. To be as precise as possible in defining rainfed and irrigated districts requires omitting these 59 districts from the analysis. On the other hand, doing so reduces the sample to 184 and thus fails to utilize valuable data. As a result, the analysis is carried out twice, once with 184 districts and again with 243. We report both sets of results in order to demonstrate their sensitivity to changes in specifications. The analysis based on 184 districts is more suitable if our intention is to compare performance of districts with distinctly different irrigation status, but the analysis based on 243 districts is more appropriate if we wish to examine trends in the aggregate, because it uses more data.

All Crops

Table 4.1 shows the growth rate of value of output for all crops for the ten zones and 3 subperiods. The calculations are based on the 184 districts whose irrigation status is constant for the period under study. The most notable output of this analysis is that the value of output grows steadily in each period, at rates of 1.75% in period 1, 2.69% in period

2, and 3.56% in period 3. For all three periods taken together, overall output grew at 2.37%; this overall growth was driven heavily by yield increases of 2.00% per year. Rising yields are a key factor in each period examined separately; in periods 2 and 3, increases in cropping intensity are also somewhat important. Increases in net cropped area are small in the first period and zero or negative thereafter. These findings are consistent with gradual increases in irrigation and adoption of HYVs and other inputs. Examining the results by agroclimatic zone shows few interesting patterns. In period 1, zone 2 (the eastern Gangetic plain) had the highest growth, followed by zones 1 (northwestern Indogangetic plain) and 3 (central/eastern highlands). In period 2, growth was fastest in zone 4 (the central/western areas of Madhya Pradesh and Rajasthan plus most of Gujarat), followed by zone 5, the south. In period 3, growth in value of output was high in all zones. When these results are broken down to account for differences in irrigation, the results change somewhat; irrigated areas of zone 1, the northwestern plains, has the highest aggregate growth overall though not the highest growth in each period examined separately.

Value of output and yields were generally higher in irrigated districts, but the difference is not great. In zone 1, the northern Indo-Gangetic Plain, value of output and yields were consistently higher in irrigated than dry districts for all three periods. This is not surprising given the high proportion of area irrigated and the low rainfall. In zone 2, the eastern Gangetic plain, and zone 3, the eastern and central highlands of Madhya Pradesh, Orissa and southern Bihar, there was no clear pattern, with only small differences between rainfed and irrigated performance. This is not very surprising given the high rainfall in these two regions. Zone 4, the central and western highlands of Madhya Pradesh and Rajasthan and most of Gujarat, had low growth in both irrigated and rainfed districts in period 1 but high growth in both thereafter. In Zone 5, covering most of the southern states of Maharashtra, Andhra Pradesh and Karnataka, output and yield grew faster in irrigated areas in period 1, but faster in rainfed areas in periods 2 and 3. We do not have an explanation for this observed pattern. As mentioned above, we also examine the results for all districts in the 450-1600 mm rainfall range, including those whose irrigation status changed during the period under study. This increases the number of districts from 184 to 243; the data are also shown in table 4.1. The results for the period as a whole do not change much between the two samples, but for some zones in some periods, the results do change significantly. As mentioned above, the 243-district sample is most useful for examining more aggregate results, so we are not concerned about differences in the disaggregated results.

All Cereals

The data for cereals, not surprisingly, show results that are more consistent with what we would expect based on our knowledge of changes in agricultural technology during the period under study. Examining the data for both the 184-district and 243-district data sets in table 4.2, growth in output and yields was highest in periods 2, the main green revolution years, and period 3, when green revolution technology was still spreading for some crops in some areas but cropped area was not growing as fast. Growth in output and yield was lowest (but still significantly positive) in period 1. Cropped area under cereals also shows expected results; it grew fastest in periods 1 and 2, when irrigation spread most rapidly, but it slowed in period 3.

Examining the data for 184 districts by zone, the fastest growth in value of output for any zone in any period comes in periods 2 in irrigated areas of zone 1, the main green revolution areas of Punjab, Haryana and Western UP, and in zone 2, the eastern Gangetic plains, in period 3. Zone 2 performs well in all periods; interestingly, its growth is driven mainly by yield increases but also by increases in cropped area. Very high growth rates of output are also found in period 1 in irrigated areas of zones 2 and 4, in period 2 in rainfed areas of zone 5 and both rainfed and irrigated districts of zones 2 and 4. Zone 3 (the central eastern plateau) performs poorly in period 1, better in irrigated areas in period 2, and fairly well in period 3 as yield grew fairly rapidly.

Individual Cereals

In this section we examine the data for 5 individual cereal crops: wheat, rice, maize, sorghum (jowar) and pearl millet (bajra). Output is measured in tons rather than prices, since physical units are comparable within a given crop. Wheat shows the most consistently high growth rates of any crop (table 4.3). In the aggregate (for the data based on both 184 and 243 districts), output growth and yield growth were high in period one but even higher in periods 2 and 3. These results are consistent with those of Byerlee (1992), who found rapid growth in wheat yields even after HYVs accounted for the vast majority of wheat acreage.

Looking at individual zones for the 184 district data set, growth in output in period 1, prior to the introduction of HYVs, was centered in irrigated areas of zone 1 (the northwest) and both rainfed and irrigated districts of zone 2 (the east). In period 2, these areas continued to have high growth in output but were also joined by irrigated districts of zone 3 (central/eastern India), both irrigated and rainfed districts of zone 4 (central/western India), and rainfed districts of zone 5 (the south). Most of these high growth rates in output continued in period 3, with the notable exception of rainfed areas of zone 5 (the south), where output fell precipitously. This is only a very minor wheat growing area, so swings in output will have little impact on nationwide performance. In all three periods, growth in output was stimulated by a combination of increases in both yield and cropped area. Rice also performed well (table 4.4), but not to the same extent as wheat. Examining the aggregate data from 184 districts, growth in value is significantly positive for all three periods; it is highest in the green revolution years of period 2 and lowest in period 3. Interestingly, period 3 shows the highest growth in rice yields, but these are more than counterbalanced by a drop in cropped area. In the data for all 243 districts, periods 2 and 3 reverse order in the ranking of overall output growth rates, while the yield and cropped area findings do not change. This shows that the rankings are not very robust; in fact, in the data for 243 districts, the three periods show nearly identical performance in growth of output. For aggregate data in which we are not concerned about changes in irrigated area, the 243 district sample is more reliable because it contains more observations.

Examining the data by zone using the 184 district sample, irrigated districts in zone 1 (the northwest) consistently show the highest growth. In period 2 this comes from rapidly increasing yields, but in period 3 it is evenly divided between yield and area growth. Notably, performance in output and yield are very poor in rainfed areas of zone 1 in period 2, the green revolution years. There is not much evidence of the rapid growth in rice output in eastern India, mentioned above, though yields do rise substantially in irrigated areas of zone 2 in period 3. On the other hand, cropped area drops greatly in this zone during the same time period. Output and yield are also high in irrigated districts of zones 4 and 5, which are mainly along the coast of the Bay of Bengal. With a few exceptions, performance is generally better in predominantly irrigated than predominantly rainfed districts. We do not have an explanation for the lack of evidence of growth in rainfed rice in recent years. There are a few notable differences between the results for the data for 243 districts vs 184 districts. In particular, output and yield performed extremely well in rainfed districts of zone 1 in period 3 when all 243 districts are analyzed. Also, the poor performance for this zone in period 2 is not found when all 243 districts are examined.

Maize shows the surprising pattern of low growth in output and yields during the green revolution years of period 2, but high growth in periods 1 and 3 (table 4.5). Yields grew at a fast rate in period one only, while cropped area grew at reasonably high rates in periods 1 and 3. One explanation for this pattern may lie in table 3.10, which shows that HYVs spread more slowly for maize than other crops. By 1984, only 33% of maize was under HYVs compared with 52% for rice and 81% for wheat. Yield growth in period one may have come from increases in irrigated area. The negative yield growth in period 2 is puzzling, however, as percent area under HYVs grew from 6% to 33%. Yields rebounded in period three to significantly positive, if not high rates. The results are robust between the 184 and 243 district data sets.

Examining the data by district, output grew at very high rates in period 1 in several zones; they were highest in rainfed areas of zones 1 (the northwest), 3 (the eastern central highlands) and 5 (the south), but irrigated zones also experienced rapid growth. Yields followed a similar pattern, and net cropped area increased significantly in most zones. In period 2, most zones experienced negative yield growth, with the exception of rainfed areas of zone 5, where it continued to grow rapidly. Irrigated areas of zone 3 and rainfed areas of zone 5 continued to show very high growth rates in output in period 2. In period 3, performance was strong in many zones. Growth in output stemmed from yield growth in some districts but area increases in others.

Only irrigated areas of zone 5 performed poorly, a result of rapid declines in both yield and area. The results for maize are mainly robust between data sets, but there are some exceptions. Sorghum (jowar) grew rapidly in both output and yields — both over 3% per year — in period 2, the green revolution years, followed by smaller increases of about 1.5% in period 3 (table 4.6). Cropped area showed practically no growth in these periods, however. This pattern is consistent with the view of sorghum as an inferior crop that will not realize increased net cropped area in response to increases in yield. In the pre-green revolution years of period 1, on the other hand, yield was stagnant but increases in net cropped area drove modest increases in output. These patterns are all insensitive to changes in the data set. Disaggregating the data does not reveal consistent patterns. Most zones show negative trends in net cropped area throughout the period; these are counterbalanced by a few zones with positive growth. In period 1, only rainfed districts of zones 1 (the northwest), 3 (the eastern central highlands) and 5 (the south) show positive growth in area, but of these, only zone 3 has positive yield growth. Only the irrigated districts of zone 4 (the western central areas) had high increases in yield, but these were more than offset by a reduction in area. Only one district had more than 1% annual growth in output. These findings are roughly constant for both data sets.

In period 2, zones 4 and 5 showed high growth rates in yield in both rainfed and irrigated areas, but they all had declining area. Nevertheless, all had positive growth rates of output. No other zones had positive output growth, and all zones had either negative or stagnant growth in net cropped area. Again, these findings are matched in both data sets. Period 3 saw mixed performance, with gains in some zones offsetting losses in others. The most significant growth occurred in yield and output in rainfed areas of zone

2 (the east), while the greatest decline was in irrigated areas of zone 5 (the south), where output fell as a result of both declining yields and cropped area.

There are some significant differences between the two data sets. For the 184 district data set, yields increased rapidly in irrigated districts of zones 1 and 5, and modest yield growth was experienced in several other zones. The high yields translated into rapid growth in output in zone 1, while several other zones had modest output growth. Cropped area rose by 8.6% per year in irrigated areas of zone 2, and slight increases in yield stimulated reasonably fast growth in output.

For the 243 district data set, on the other hand, no district had more than 1.7% output growth, and some had highly negative growth in output. All the zones that experienced high yield gains had negative growth in area. The difference between the two sets of output suggest that the zones that shifted from under 25% irrigated to over 25% irrigated between the two periods generally had higher yield growth but a greater reduction in area, resulting in an overall smaller increase in output. Again, this is consistent with the performance of an inferior cereal — the market is not big enough to absorb increases in yield, so area falls.

Bajra (pearl millet) might be expected to have a similar pattern to that of sorghum, but it does not. Table 4.7 shows that rapid output growth was driven by yield increases in period 1, after which stagnant output in period 2 resulted when yield increases roughly countered declining area, and then rapid growth in period 3 resulted from modest growth in both area and yields. Overall growth in both output and yields was just under 2% for the period as a whole, while area was about constant. The general direction of these findings is the same under both data sets, though the specific numbers differ somewhat; growth rates are significantly higher in period 1 but significantly lower in period 3 when all 243 districts are analyzed, and overall growth rates are slightly higher.

Positive growth in output in period 1 resulted from a mixture of increases and declines in area and yield in different zones. Three zones had an increase in both: these include rainfed districts of zone 1 (the northwest), irrigated districts of zone 3 (the central/eastern highlands), and irrigated districts of zone 4 (the western/central areas). Zones 1 and 4 had very rapid output growth. Several other zones had yield and area moving in opposite directions, and rainfed areas of zone 5 (the south) had negative growth in area, yield and output. These patterns are largely duplicated in the data for 243 districts, though output changes from stagnant to 2% growth in rainfed areas of zone 4.

In period 2, irrigated districts of zones 3 and 5 show positive growth in output, but other zones have either negative or stagnant output. Most districts either declined in both yield and area or one or the other. Rainfed areas of zone 1 showed a very sharp drop in yields and output. These findings are roughly constant across data sets. In period 3, the poor performance of rainfed areas of zone 1 were completely reversed, with rapid growth in yield, area and output. High output growth in irrigated areas of zone 2 is driven by area increases, and output growth in irrigated areas of zone 4 is driven by growth in both area and yield.

Most zones show rapid yield growth, while growth in cropped area is significantly positive in some areas but significantly negative in others. These findings are qualitatively the same in both data sets, but magnitudes of some indicators change significantly between one district and another. Irrigated districts of zone 4, for example, show only 2.3% growth in output in period 4 in the 243 district data set compared to 10.6% in the 184 district data set.

Summary Comments on the Tabular Analysis

Growth rates of the value of output, yield, cropped area and cropping intensities provide an indication of the pattern of agricultural growth under different conditions in India. On the whole, the findings reflect our prior expectations; for example, irrigated areas in zone 1, the northwestern green revolution belt, consistently show the highest growth rates in yield and output, particularly in periods 2 and 3.

Contrary to our expectations, we did not find much evidence of exceptional rice and sorghum growth in period 3 in favorable rainfed areas eastern and central India. However, in most zones for most crops, output and yield in predominantly rainfed districts grew quite rapidly, nearly as much as in irrigated zones. This suggests that rainfed agriculture has performed quite well, even if not up to the high standards set by the performance of irrigated agriculture.

It is important to reiterate that the findings of the tabular analysis are more indicative than definitive of trends in irrigated and rainfed agriculture in different agroecological zones of India. As mentioned above, sensitivity of the calculated growth rates to specification of the time periods and zones requires that we treat these results with caution. More precise understanding requires a more detailed analysis that could not be undertaken with the resources available for this study.

PRODUCTION FUNCTION ANALYSIS OF SOURCES OF GROWTH IN PRODUCTIVITY

We use the district data to estimate a production function for Indian agriculture. As mentioned earlier, the data cover 243 districts, or most of those in the Indo-Gangetic plains and peninsular India with an average annual rainfall between 450 mm and 1600 mm. Those with desert conditions (as in Western Rajasthan) or very high rainfall (as in the Western Ghats) are excluded, since their conditions are not comparable to those in the rest of the sampled districts.

The production function approach has some important advantages over tabular analysis. First, there is no need to define districts as either irrigated or dry; instead, each district is associated with a continuous variable indicating the percent irrigated area. Second, there is no need to divide the sample into subperiods. Third, we can classify districts into disaggregated agroclimatic zones without encountering presentation difficulties, as each zone simply adds an additional line to the output.

The data provide us with sufficient degrees freedom and enough districts in each zone to categorize all districts according to ICAR's 20- See section 2 for a listing of the 20 zones. Many districts straddle two zones, in which case we put them in the zone in which most of the district lies.

Coastal districts in Tamil Nadu, Orissa and Andhra Pradesh lie partially in zone 18 and partly in adjoining zones 7, 8, 12 and 15; in this categorization they are all placed in the adjoining zones. Zone 3 is excluded from the production function analysis because it contains only one full district; others are placed in adjoining zones. zone system. The zones included in our sample are 2, 4, 5, 6, 7, 8, 9, 10, 11, 12, 13, and 15, for a total of 12 of the 20 zones.5

The production function is a Cobb-Douglas type in which the coefficients represent elasticities; the dependent variable is the value of output. Most of the variables in the production function are self-explanatory. They are listed in table 4.8.

Prices in the analysis are taken as the average of prices prevailing over the overall period under study. Constant prices are taken to avoid terms of trade effects that change during the course of the study period. In a subsequent analysis a more sophisticated method will be used to deflate prices over the course of the study period, but the system used here is sufficient for our purpose, which is to obtain a preliminary understanding of the contribution of various factors to agricultural production.

Table 4.8 Explanatory variables in the production function analysis of sources of growth in value of output

Variable	Description
LOGGCA	log of gross cropped area
LOGPIAGC	log of percent of gross cropped area that is irrigated
LOGLAB	log of number of labor days
LOGFERT	log of tons of fertilizer used
LOGBULL	log of number of bullocks
LOGTRAC	log of number of tractors
LOGPHYV	log of percent area under HYVs
LOGMKT	log of number of regulated markets
LOGLITE	log of adult literacy rate
LOGROAD	log of distance of paved roads divided by gross cropped area
LOGRAIN	log of annual rainfall (mm)
LNLAGTT2	log of lagged terms of trade
DROUGHT	dummy variable; 1 if district is drought prone, zero otherwise
TREND	year
Z2...Z15	dummy variables indicating agroclimatic zones from ICAR's 20- zone classification*

* Zone 2 is the reference case in the analysis.

The estimated production function for 243 districts is presented in table 4.9. The model's explanatory power is reasonably high. All variables except the logarithm of literacy and the dummy variables for zones 10 and 11 are significant at the 10% confidence level; the remainder are all significant at the 1% confidence level.

Table 4.9 Coefficients of the production function variables

	Variable	Coeffecient
INTERCEP	-1.77*	(0.174)
LOGGCA	0.72*	(0.014)
LOGPIAGC	0.13*	(0.016)
LOGLAB	0.12*	(0.014)
LOGFERT	0.05*	(0.004)
LOGBULL	0.01*	(0.003)
LOGTRAC	0.12*	(0.004)
LOGPHYV	0.02*	(0.004)
LOGMKT	-0.02*	(0.005)
LOGLITE	-0.01	(0.015)
LOGROAD	0.03*	(0.009)
LOGRAIN	0.16*	(0.013)
LNLAGTT2	0.09*	(0.017)
DROUGHT	0.07*	(0.010)
TREND	0.01*	(0.001)
Z4	0.25*	(0.032)
Z5	0.32*	(0.033)
Z6	0.14*	(0.035)
Z7	0.14*	(0.037)
Z8	0.30*	(0.038)
Z9	0.28*	(0.036)
Z10	0.05	(0.034)
Z11	0.01	(0.039)
Z12	0.37*	(0.036)
Z13	0.16*	(0.040)
Z15	1.26*	(0.042)

R-square = 0.77

Note: Standard errors in parentheses

On the whole, the elasticities are quite small, in most cases less than 0.15%. The regression coefficients sum to about 1.15, suggesting the presence of economies of scale. The variables with the highest values are gross cropped area, percent irrigated area, rainfall, labor, tractors, and several of the zone dummies.

Most of these results are not surprising; in a country where water is scarce in most places most of the year, it makes sense that rainfall and irrigation have a large impact on value of production.

Gross cropped area has the largest coefficient because the dependent variable is the total value of output, not the value of output per hectare. The positive zone dummies are also to be expected, because the base zone is 2, on the fringe of the Rajasthan desert where output is expected to be low. The strong positive contribution of tractors is not entirely straightforward.

It may in fact reflect the importance of overcoming power constraints in land preparation. A possible alternative is that tractors act as a proxy for other variables related to infrastructure and capital that contribute positively to the value of output. Yet another possibility is that tractors may be highly collinear with bullocks and labor, two other variables that represent a "horsepower" input. If so, tractors may simply capture much of the effect resulting from those two variables. This would be consistent with the strong positive contribution of labor.

The significantly negative value of the coefficient for markets is puzzling, as is the negative but insignificant coefficient for literacy. The negative coefficient for markets represents the direct effect of the number of government regulated markets on output.

However, there may also be indirect effects that this analysis does not attempt to identify; for example, markets may increase input use, thus increasing production *indirectly* despite the negative *direct* effect in the production function estimation.

More detailed analysis is needed to fully explain the counterintuitive result shown here. A simultaneous equations model would be needed to capture any possible indirect effects and eliminate endogeneity.

The negative value of the lagged terms of trade is also somewhat difficult to explain. This finding suggests that the value of agricultural output declines after agricultural prices rise relative to nonagricultural prices.

A reasonable explanation for this result would be that demand declines in response to higher prices, thus reducing the value of output in the subsequent period. However, agricultural supply is relatively inelastic, and terms of trade generally do not move very much over time (Hazell et al 1995).

A possible alternate explanation of the negative coefficient of the terms of trade would be reverse causality: that terms of trade respond negatively to changes in output, which would make sense. Yet another alternate explanation is the same as that for the negative coefficient for markets, explained above: terms of trade may have a negative direct effect on the value of output, but a positive indirect effect through positive effects on input use.

Higher terms of trade would reduce the price of inputs relative to outputs, presumably increasing their use and thus increasing production. Again, in this analysis we cannot measure any indirect effects that might exist.

The coefficients of the zone dummies indicate the change in the intercept for each zone after controlling for all the other variables, such as irrigation, rainfall, and technology adoption. Remaining factors that the zone dummies may capture include soil, sunlight and slope conditions, infrastructure, factor markets conditions, and various state-level policies and institutions. By far the largest coefficient (1.26) is for zone 15, which covers the Gangetic plains of West Bengal.

Zones with the next highest coefficients, ranging from 0.25 to 0.37 are 4 (Northwest Indo-Gangetic plains and uplands areas), 5 (plains and highlands areas of eastern Gujarat and western Madhya Pradesh), 8 (Tamil Nadu, southern Karnataka, and Chittoor district of Andhra Pradesh), 9 (northwestern Indo-Gangetic plains or Punjab, Haryana and Uttar Pradesh), and 12 (Orissa, Bastar district of Madhya Pradesh, and north Coastal Andhra Pradesh). A third tier of zones with coefficients ranging from 0.14 to 0.16 are 6 (Maharashtra and northern Karnataka), 7 (Andhra Pradesh and central Madhya Pradesh), and 13 (eastern Gangetic plains areas of Bihar and Uttar Pradesh).

Other zones (10 and 11, representing, respectively (western Madhya Pradesh and the uplands of eastern Madhya Pradesh and southwestern Bihar) have small positive coefficients.

While the specific values of the zone dummy coefficients may be difficult to interpret, the signs and most of the ordinal rankings are not.

The Gangetic plains areas have the best soils and other growing conditions, for example, both West Bengal and Gujarat have strong infrastructure and service organizations supporting agriculture, so these areas should be expected to have strong performance even after accounting for the other variables in the production function.

As we discuss later, in section 7, West Bengal has the strongest local government institutions. Other zones are not expected to so well as they face inferior soils and weather (much of the south) or are in economically undeveloped areas with poor infrastructure (the central and eastern highlands of Madhya Pradesh, Orissa and Bihar). In a subsequent study, the production function analysis will be expanded to include a more detailed decomposition of the sources of growth of output.

Like the tabular analysis, the estimation presented here gives an indication of the sources of agricultural output if not the detailed information needed to design future agricultural development strategies. The relatively large, significant coefficient of percent irrigated area confirms the need to construct an agricultural typology that accounts for differences in irrigation. More detailed analysis may suggest additional criteria for defining typologies.

5

TECHNOLOGICAL CHALLENGES IN RAINFED SYSTEMS

The immense diversity of rainfed areas in India defies easy generalizations. Nonetheless, in most rainfed areas the main challenges to developing and diffusing new technologies revolve around soil and water management (exceptions may be some of the rainfed areas in the Himalayan valleys).

In most areas, conservation of soil moisture, both in situ and ex situ, is the major challenge. However, in high rainfall areas, and even in many medium rainfall areas with heavy soils, excess moisture in the monsoon season is also an important problem in many years.

In general, rainfed agriculture is characterized by substantial heterogeneity over time and space. Season-to-season variation in the amount and timing of rainfall is a major challenge to crop management and the applicability of new technologies. Crop yields are highly variable over years, as shown in the previous section.

Perhaps as important or more important than variability over time is spatial variability across fields and farms due to microclimatic differences in soil type, topography and irrigation status. For example, interfarm variation in yields in predominantly rainfed villages is often greater than 50%, compared to 20-30% in irrigated areas (Byerlee and Hussain, 1992; Shah and Sah, 1992).

These microlevel variations have often been found to be major factors explaining adoption of new technologies. Studies which ignore these variations often mistakenly blame the resource situation of farmers for low adoption (Gupta, 1991; Shah and Sah 1992). In some regions, these variations can result in a bewildering array of microecosystems each with their own form of management (e.g., see Mahapatra, 1990 for a description of rainfed rice ecosystems in Eastern India).

A third aspect that characterizes much rainfed agriculture is the close interaction of crop and livestock production. Crop residues are a key source of animal nutrition; in rainfed areas crop residues may constitute over half of the value of crop production.

On the other side, livestock provide the bulk of draft power and manure for soil fertility improvement in rainfed areas.

However, since fodder supplies are limited, draft power is often inadequate at the beginning of the monsoon, limiting both the intensity and timeliness of farm operations. Further system interactions are involved in the management of rangeland and forest resources, often as common property of the community.

Since livestock range freely in the dry season, farmers cannot make individual decisions on crop intensification or range management. The decline in availability of fuelwood from community lands means more reliance on manures for household fuel, reducing the amount returned to the soil. Thus technology interventions in rainfed areas usually have to consider the whole system, including crops and livestock, and farm and as well as community management.

ISSUES IN TECHNOLOGY DEVELOPMENT

Seed Technology

1. From the information on the spread of HYVs presented in section 3, it is clear that improved seeds have been the most important source of productivity growth in rainfed areas over the past decade or more (Walker, 1989; Ryan and Walker, 1990). The success of improved seeds is due to several factors:
   - The development of early maturing and pest resistant varieties and hybrids
   - The release of improved varieties for different agroecological systems.
   - The improved distribution of seed, especially with the increased participation of private seed companies

Nonetheless, there remain significant challenges to the further spread of HYVs, especially since the areas where adoption has still not taken place are the more marginal areas. 1. A key factor in farmers' acceptance of new varieties in the drier areas is the increasing importance of crop residues as livestock feed. Over time, the price of fodder has risen in real terms in rainfed areas and in many areas half or more of the value of cereal crop production is from fodder (Kelley et al., 1993). Despite the importance of fodder, breeders have historically focused on grain yield. Kelley et al. present evidence that there are significant differences in the yield and quality of fodder by variety. In dry areas, traditional varieties are often preferred for their higher fodder yields and because of a price premium for quality.

2. Grain quality is also an important consideration in many rainfed areas where farmers specialize in producing crops that command a price premium. Examples include postrainy season sorghum and

durum wheat where rainfed farmers may receive price premiums of 50-100% for quality produce. These quality premiums provide an additional challenge to crop breeders and help explain the low adoption of HYVs in some areas.

3. There are also many "hard core" cases where farmers produce crops under very harsh conditions and where it will be very difficult for breeders to make significant breakthroughs. These include the crops produced under post-rainy season conditions of receding moisture (e.g., sorghum, wheat and gram), millet in sandy soils and rice in areas of poor water control (e.g., rice in upland systems or the "beushening" system).

There are clearly opportunities for further expansion of improved varieties in rainfed areas. The largest such opportunity is through wider use of improved varieties of pulses where the current adoption rate is still very low. Recently released varieties of gram will probably have significant impacts in the next few years (Kelley and Parthasarathy Rao, 1994). With closer attention to farmers varietal needs in terms of maturity, fodder, grain quality etc., HYVs will likely continue to spread in rainfed areas. Given the diversity of rainfed ecologies, some have argued for more decentralization of crop breeding programs and greater participation by farmers in varietal selection (Maurya et al., 1989). The basis of this argument is that it will be difficult to develop widely adapted varieties for highly diverse systems. However, there are many examples of a variety that has been widely adopted across a wide area in rainfed agriculture—for example, the sowing of wheat variety C306 over millions of hectares in Central India. On the other hand, the extreme heterogeneity of rice ecologies in a small area each with its own traditional varieties may require a different breeding strategy. Fertilizer Use Despite the growing use of fertilizer in rainfed areas, soil fertility is a major problem. Use of organic manures is inadequate and many plots do not receive fertilizer or manure over a period of several years of continual cropping (Desai and Rustagi, 1994.). Chemical fertilizer is still only used on a little over half of the rainfed grossed cropped area. Use of fertilizer is related to rainfall, access to credit and proximity to a town (Kumar and Desai, 1994). Use of HYVs and an efficient fertilizer distribution system may partly compensate for lower response to fertilizer in drier areas as shown by the example of Gujarat where most farmers use modest doses of fertilizer even in dry conditions (Nampootheri and Desai, 1995; Desai, 1985). Fertilizer use is also generally much more widespread on cash crops compared to food grains as shown by Table 3.5 above (NCAER 1990). Indeed it is frequently observed that fertilizer use varies considerable within a farm, with many farmers applying fertilizer to a cash crop such as cotton but not to a food crop. This of course, reflects the relative profitability of fertilizer use and the fact that fertilizer use on food grains is often not very profitable (Ranjaswamy, 1990). Thus in many areas, farmers are already familiar with fertilizer use. The relatively slow rate of intensification of fertilizer use (measured by the rate of fertilizer use per fertilized ha) also suggests that further gains will be made through expansion of fertilizer to areas that do not currently receive fertilizer. Since these are generally areas characterized by severe drought stress or poor water control, future progress in fertilizer use in rainfed areas is likely to have lower marginal gains than in the past.

In dry areas too, the profitable use of fertilizer will generally require greater fertilizer efficiency through a balanced dose, timely planting, time of application and placement. It is also likely that fertilizer will only be profitable in marginal areas in some years and locations

(Dvorak, 1992). This suggests a movement away from general recommendations to more specific recommendations conditional on factors such as crop rotation, moisture availability, and time of planting. In addition to the important agroclimatic effects on fertilizer response, other demand related factors affecting fertilizer use are farmers' knowledge and access to credit and use of improved varieties. Recent studies have shown that supply-related factors also influence fertilizer use, especially access to a town. Over the next decade or so, fertilizer use in rainfed areas is likely to continue to steadily expand as these various supply and demand related constraints are relaxed. Probably the major intervention needed to accelerate the speed of diffusion is more location specific research to generate conditional recommendations, combined with a change in extension focus away from the package approach toward more specific information, and increasing farmers' understanding of factors affecting fertilizer efficiency.

Soil and Water Management

Although farmers have practiced various forms of soil and water management (SWM) for centuries in rainfed India, there is surprisingly little study of the extent and effectiveness of these traditional SWM systems. Some systems such as the Haveli system in Madhya Pradesh to harvest water in the kharif season for rabi planting have been widely adopted, as has terracing of steep land especially for rice cultivation. These systems are also quite effective in improving soil and water management. Other traditional systems such as border bunds are also widely used but are less effective in SWM, especially soil erosion control. However, farmers have multiple objectives for adopting SWC technologies (e.g., demarcation of fields, ease of land preparation etc.) and these traditional systems often meet these multiple objectives better than various introduced systems (Kerr and Sanghi, 1992). Since the 1920s, considerable efforts have been made to develop and extend SWM technologies in rainfed areas of India. These can be summarized in several stages:

1. Early work in pre-independent India to introduce essentially engineering approaches to soil and water management—especially the Bombay Dryland Farming method. These methods which focused on contouring and other land improvement techniques, gave little emphasis to agronomic and institutional issues in increasing productivity. As a result, impacts on crop yields were very modest and adoption was low (Singh, Vijayalakshmi, Sullivan and Shaw 1987, Ranjaswamy 1990).
2. Beginning around 1970, renewed attempts were made to develop packages for dryland farming, especially through the newly established all India dryland projects and ICRISAT. These packages, based on a microwatershed approach, differed from their predecessors in including both techniques for SWC as well as improved cropping systems and agronomic practices. Considerable attention was also given to in situ soil and water management, through improved tillage and drainage. Overall these packages promised yield increases of three to four times the traditional methods. However, adoption of these packages has been low, although elements of the packages, especially the agronomic practices, have been fairly widely adopted in some states. Adoption of SWM components of the packages have been very slow (Walker and Ryan, 1990; Kshirsagar and Ghodake; Kerr and Sanghi, 1992). The relatively high cost of the package, draft power constraints, and lack of community organization all contributed to the low uptake of these technologies.

3. In the 1980s attention shifted to integrated watershed management (IWM) projects that combined elements of the microwatershed approach with efforts to manage the whole watershed including community lands. Watershed projects are often quite complex, including components for water harvesting, forestry, engineering works, improved agronomic practices etc. Although total costs were low in relation to major irrigation works, a large part of these costs were in the form of subsidized engineering works for land improvement and inputs. Watershed management projects are discussed in more detail below.

Protective Irrigation and Water Harvesting

Given that water is the limiting natural resource in agricultural production, scientists and policy analysts have long sought to develop cost effective mechanisms to secure one or two protective irrigations for dryland crops.

This idea is very attractive given the substantial within year rainfall variability in much of India. Moisture stress is an annual threat in SAT areas, where 2 to 3 week dry spells are common during the rainy season.

Two approaches to protective irrigation are harvesting runoff water to store for subsequent irrigation purposes, and extensive irrigation from existing irrigation sources. Water harvesting is a common component of watershed projects (described further below).

The idea is to channel runoff from agricultural plots into small farm ponds, where it is stored long enough to provide irrigation water during a dry spell or in the postrainy season. The principle is similar to that of the traditional irrigation tank, except on a much smaller scale.

Simulation studies conducted at ICRISAT, however, found that the costs of water harvesting were unlikely to exceed the benefits in semi-arid areas (Walker and Ryan, 1990). This is because water is most likely to be available in the pond when crops enjoy abundant moisture, but when rain is sparse and crops suffer from moisture stress, the pond is likely to be empty. This state of affairs could be altered by costly investments to line ponds with plastic or cement that prevents percolation, but the financial costs would exceed the returns.

Further studies at ICRISAT found that water harvesting might be cost effective under certain circumstances in higher rainfall zones in central Madhya Pradesh with moistureretaining black soils. In these areas, studies suggest that water harvesting toward the end of the rainy season could provide enough moisture at the start of the postrainy season to support a postrainy season crop grown on residual moisture, after the harvest of the rainy season soyabean crop.

Water stored in a farm pond is likely to have a significant impact on postrainy season crop growth in two out of every three years, which might be enough to make it costeffective (Pandey 1986). It is worth experimenting with small scale water harvesting systems in relatively high rainfall areas.

Protective irrigation from canals and wells is another possible means of making available water go farther. Dhawan (1988b) argues that it is more realistic to increase agricultural output by using available irrigation more prudently than by increasing the yields of purely rainfed crops.

In particular, he cites evidence that yields of many rainfed food crops can be boosted significantly with one or two protective irrigations to supplement the moisture supplied by rainfall. In many water-scarce regions, however, irrigation water is used intensively for such crops as paddy, sugarcane and horticultural crops, while no water is allocated to rainfed crops. There is scattered evidence of farmers shifting to crops that require only a few irrigations when the water supply is unreliable. (Kerr 1993) found that farmers in a village in Andhra Pradesh shifted from paddy (which requires daily irrigation) to groundnut (which requires weekly irrigation) when electricity supply became subject to 6Meri Whitaker, Economist, and Eva Weltzien, Millet Breeder, ICRISAT, unscheduled power cuts during the summer season. Whitaker and Weltzien6 (personal communication) indicate that farmers in dry areas in Rajasthan or on the fringes of irrigated areas irrigate millet extensively when they are not sure how much water they will have access to.

However, substantial irrigation resources remain allocated to water intensive crops. Some of the reasons for this behavior are quite obvious; for example, sugarcane is popular in Maharashtra because it is easy to manage and fetches a high price.

Paddy is popular in Andhra Pradesh because it is the staple food grain. Presumably, if price signals indicated a growing scarcity of food crops, irrigation resources would probably shift endogenously toward more extensive use. However, it is worth examining the factors that would affect such a decision. Further work is needed to understand the private and social costs and benefits of extensive vs intensive irrigation, the circumstances under which farmers practice one as opposed to the other, and policy tools that can be taken to encourage the most efficient use of irrigation water.

ISSUES FOR THE FUTURE

How Much Research?

The above review suggests that the impacts of agricultural research in rainfed areas have been uneven, with significant successes in some areas and types of technologies and almost no impacts in other areas and technology types. As a broad generalization, we can say that varietal improvement research has had major impacts over the past two decades in most crops, while research on SWM has had little impact. One logical conclusion often reached from such an observation is that the more modest successes of research in rainfed areas reflects the bias in allocating research resources toward irrigated areas.

In the case of crop improvement research, there is little evidence of such a bias. Research intensities based on ICAR allocations are generally higher for rainfed crops than they are for rice and wheat (World Bank, 1990) (Table 5.1). Recent work by Mruthyunjaya et al (1995) confirms these results. In addition, for some rainfed crops, especially maize, sorghum, millet, cotton and some oilseeds, public sector investments in research are matched by private sector investments (Pray et al., 1991; Singh et al., 1995). Even within a commodity, there is no evidence of systematic underinvestment in crop improvement research for rainfed areas, as shown by the recent analysis of allocation of research resources to wheat improvement research where rainfed areas receive a high share of resources relative to their contribution to value of production (Jain, Byerlee and Traxler, 1996).

In addition, the area of some rainfed crops is declining over time due to low demand (e.g., sorghum and millet) or conversion of land from rainfed to irrigated, requiring a decrease in the share of research resources invested in these crops. Unfortunately there are no comparable data on the share of resources being allocated to SWM research for rainfed areas. There are significant investments through CRIDA and the all-India coordinated project for rainfed areas, but there is no baseline to show to what extent there may be systematic under or over-investment in SWM research for rainfed areas.

Table 5.1 Total (plan and non-plan) ICAR expenditure in current rupees by commodity1

	6th Plan (1980 -85)		7th Plan (1986 - 90)	
	Annual Avg. Exp. (Rs M)	Share in Total Exp.	Annual Avg. Exp. (Rs M)	Share in Total Exp.
Food Grains	64.9	0.11	107.1	0.11
Rice	31.6	0.05	50.0	0.05
Wheat	11.7	0.02	19.1	0.02
Barley	3.6	*	5.9	*
Maize	7.6	0.01	6.2	*
Millet	5.7	0.01	18.0	0.02
Sorghum	4.7	*	7.9	0.01
Pulses	19.7	0.03	44.4	0.04
Oilseeds	24.4	0.04	45.3	0.04
Forage Crops	13.0	0.02	24.1	0.02
Cash Crops	108.7	0.18	167.3	0.17
Sugarcane	20.0	0.03	31.9	0.03
Sugar Beet	0.0	0.00	1.0	*
Cotton	26.7	0.04	41.1	0.04
Jute	17.3	0.03	29.2	0.03
Tobacco	17.9	0.03	23.0	0.02
Plantation Crops	26.8	0.04	41.1	0.04
Horticulture Crops	78.7	0.13	115.8	0.11
Fruits & Vegetables	49.4	0.08	69.5	0.07
Tubers	5.7	0.01	10.1	0.01
Potato	19.1	0.03	28.8	0.03
Floriculture	4.0	*	5.7	*
Mushroom	0.5	*	1.7	*
Crop Total	308.4	0.52	504.0	0.50
Animal Sciences	170.7	0.30	289.2	0.29

Bovines &				
Large Animals	127.6	0.22	216.9	0.22
Small Stock	25.9	0.04	43.5	0.04
Poultlry	13.6	0.02	21.8	0.11
Other	3.6	*	7.0	
Fisheries	74.6	0.13	117.0	0.12
Soils, Etc.	34.5	0.06	95.0	0.09
Total	589.2	1.00	1,005.2	1.00

1 ICAR, including AICRP expenditures can be disaggregated by commodity because most institutesand AICRPs are devoted to only one or a few crops. IARI expenditure was allocated according to the number of scientists working on each crop. Disaggregated SAU research expenditure is not available. As state governments provide 25% and ICAR 75% of AICRP costs ICAR expenditures on AICRPs were increased by one third. This still leaves out a large part of state research expenditure. * Less than 0.01.

Source: ICAR unpublished computer printouts.

Approaches to Research

Tapping existing potential vs. developing new potential. Productivity increases can be achieved both by encouraging farmers to attain yields that are achievable with existing technology (closing the yield gap between research stations and farmers' fields) and developing new technologies that increase yield potential (widening the yield gap in the very short term).

These two approaches are not mutually exclusive, but rather can be carried out side-by-side. Clearly, both approaches are important. On the whole, closing the yield gap requires working relatively closely with farmers to identify the causes of their yield shortfalls, while expanding the yield ceiling involves work mainly on research stations. Indian agricultural research tends to focus heavily on on-station work compared to on-farm work, suggesting that there are potential gains to be made from marginal increases in the allocation of resources to on-farm work.

Table 5.2 Yield differentials among the selected rainfed and irrigated crops

Crops and Variety	Yield (Kg/Ha)			
	On Experiment Farms1	District Average (1988-89)	Yield Difference	Untapped Potential (col.3 as % of col.1)
Rainfed Crops				
Bajri (BJ 104)	2200	1155	1045	48
Maize (G1)	2870	1413	1457	51
Cotton (Khapatio)	848	143	705	83
Groundnut (GAUG 1)	1480	1249[3]	231	16
Til (G 1)	630	603	27	04
Irrigated Crops				
Paddy (GR 138)	4580	2153	2427	53
Wheat (Lokvan)	3980	2627	1353	34
Cotton (H 6)	1336	235	1102	82
Tobacco (Calcutti)	3040	1672	1368	45
Sugarcane	9500	8603	897	09

1 Based on the average yield obtained through the research experiments (pooled over time and location).

2 As in Table 4 except for maize.

3 Average for 1988-89 and 1989-90 for, 1988-89 was an exceptionally good year for groundnut yield.

Source: Director Research, Agricultural University, Ahmedabad.

Shah and Sah (1993) present several types of evidence to support this contention. Table 5.2 shows their assessment of differences between on-station and on-farm yields; they point out that it is unlikely that this gap could ever be closed completely because farmers do not enjoy the idealized conditions on research stations.

Table 5.3, however, shows interfarm yield variations and the gap between the most productive farmers and their more average neighbors.

Table 5.3 Dryland farming: Interfarm variations in yield

Crops	Yield (Kg/Ha)			
	Maximum	Average of Top 10% of Farmers	Average of all Farmers	Co-efficient of Variation
Rainfed1				
Bajri	2478	1809	895	54
Groundnut	1486	897	3842	62
Cotton	1651	1007	443	54
Irrigated[3]				
Wheat	5005	4956	3641	24
Paddy	7136	6740	4338	34
Mustard	3023	2726	1778	31
Castor	4956	4423	2455	51
Cotton	1982	1982	1343	37

1Based on the sample survey of farmers in the dryland region (see Shah and Sah, 1991).

2The year was particularly bad in the survey area.

3Based on the sample survey of farmers in the irrigated region (see Sah and Shah, 1992); the sample was selected from the adjoining villages unlike that in the other survey of irrigated crops for the study on Soil Testing Services (Sah and Shah, 1990).

The tables show two interesting points; first, yield gaps are much lower in irrigated crops than dryland crops, and second, more interestingly, the gap between the top 10% of farmer and the average farmers is almost as high as that between the average farm and the research station. Taken at face value, this finding suggests potentially high returns to helping the average farmer become more like the high performing farmer. Of course, much of the difference in yields may result from variations in soil conditions that cannot be overcome, but it is likely that variations in management also are important.

Research trials in agronomy and soil and water management routinely test an "improved" management system in relation to "the farmers' practice," as if all farmers managed their fields identically. Generally "the farmers' practice" resembles that of the average farmer, and no comparison is made between the "improved" system and that of the best farmers. Performance gaps might be found to be considerably smaller if research station results were compared to those of the top farmers.

Of course, the government extension system is designed to help spread technologies and management practices so that average farmers improve their performance. Extension services, however, operate on the principle of taking findings directly from the research station to the farm. No emphasis is put on helping the average farmer learn lessons from the best farmer.

Some NGOs, notably the Aga Khan Rural Support Programme (AKRSP) in Gujarat, have developed farmer-to-farmer extension programs designed explicitly to reduce interfarm variations in performance by helping average farmers learn from better farmers. One of the principles implicit in this approach is that even top-performing farmers operate under conditions more similar to those of the average farmer than those of the research station, so they may have more to offer to average farmers than would the extension worker who brings knowledge from the research station.

Impressive results have been achieved from this approach; Shah and Kaul Shah (1994) report production increases of 30%-100% in the project villages they study, though these gains result only partly from farmer-to-farmer extension and partly from other sources. Farmer-to-farmer extension is also spreading in other countries with growing success.

Need for farmer-based research. This review of the research directed to rainfed areas indicates that much of the research has been experiment-station based using promising SWM techniques as the point of departure for developing technological packages for rainfed areas. In particular, there has been relatively little in depth research to understand traditional systems and farmers' rationale for following particular practices. Even the ICRISAT village studies, while providing an excellent analysis of how factor markets work at the village level, devoted relatively little attention to describing and understanding traditional farming systems.

Where in-depth efforts have been made to understand local systems, the results often suggest that much of the technology being promoted to farmers is not relevant to their agroecological and socioeconomic circumstances. Excellent examples include the careful studies of the Bueshening system for rainfed rice (Fujisaka et al, 1991, Singh et al., 1994), which show that practices used by farmers, such as high seed rates and ploughing of seedlings, provide similar yield to recommended systems at lower costs and have other advantages such as more dispersed seasonal labor use patterns, greater flexibility to withstand drought and floods, and more yield stability.

Similarly, Kerr and Sanghi (1992) describe farmers' traditional SWM practices and the rationale for their use. Once understood in terms of farmers' objectives and resources, the reasons for the low adoption rate of recommended SWM practices become obvious. Likewise, detailed farm level studies in dryland Gujarat help identify the constraints on soil fertility management under conditions of severe water stress (Shah and Sah 1993). These examples indicate that the focus of research on SWM should now shift from the research station to the farmers' fields with farmers' active participation.

This implies the need for detailed farm-level diagnostic studies to understand existing systems and practices in different agroecological zones, combined with onfarm testing of promising technological components identified in the diagnostic studies. It is also important that this work be closely linked with work on the experiment station to ensure that findings from the onfarm work be fed into the design of on-station research.

Such an approach to research is in its infancy in India, but it has shown that it can yield valuable results. Pimbert (1991), for example, used matrix ranking and other participatory rural appraisal (PRA) approaches to learn the preferences of groups of Indian women farmers. Matrix ranking had the advantage of displaying the problem at hand in a way that was easy for both farmers and researchers to understand. The women indicated numerous uses of pigeonpeas, including grain to consume at home, grain to sell, leaves to use as fodder, and stalks to use as construction material. They also listed various preferred characteristics, such as seed yield, market price, pest resistance, storability, taste, and yield and quality of leaves and stalks. Using matrix ranking, they indicated the relative importance of each of these characteristics. They also explained that usually they plant more than one variety in order to meet their various objectives.

Based on the information thus collected, Pimbert searched computerized databases to identify varieties with characteristics likely to be attractive to the women farmers. He found that some varieties that had been rejected by researchers had traits that were likely to make them attractive to farmers. He then offered several varieties to the farmers, who planted small amounts of each, including some local varieties, a variety that recently had been officially released, and some improved varieties that had not been released. At the end of the season, the women again used matrix ranking to rate the performance of each variety in terms of all the criteria identified previously (figure 5.1). They found three unreleased varieties to meet their various needs, but they unanimously rejected the officially released variety due to its bitter taste.

Pimbert's study shows the power of participatory research approaches to help guide scientists to design technology that meets clients' needs. In particular, he found that scientists' traditional focus on seed yield as the sole evaluation criterion led them to design an unacceptable technology. Using information gained through participatory research, subsequently they could reallocate their research resources more effectively. Similar work is underway in collaboration between ICRISAT, ICAR and the Government of Rajasthan to involve farmers in the selection of pearl millet varieties for marginal environments. Sanghi (1989) reports initial efforts at participatory research in dryland agriculture; Gupta (1991) and Sanghi et al (1994) present frameworks for organizing and institutionalizing participatory work in the future.

Finally, it is important to point out that significant expansion of on-farm, participatory research will require significant changes in the culture of agricultural research in India. To date, on-farm research has been limited almost entirely to demonstration trials in which a previous experiment is simply duplicated on a farmer's field. Farmers have no little or no input into these scientist-dominated exhibitions. Gupta (1989), arguing in favor of participatory on-farm research, documents agricultural scientists' attitudes that farmers' indigenous practices are not worthy of study. Sanghi et al (1994) note that while farmers rejected recommended soil and water conservation practices because they were not suitable to existing farming systems, soil conservation officials and scientists believed the reason was that farmers did not understand nor care about erosion. Such attitudes and approaches will have to change if progress is to be made.

Package vs gradient approach to technology transfer. Closely related to the above challenge is the fixation of the research system on developing technological packages for rainfed areas. These packages have become steadily more complex over time as the strategy has moved from SWM to include agronomic practices and various components of integrated watershed programs. Yet packages are of dubious relevance in rainfed areas due to (1) the diversity in time and space of production conditions, (2) risk factors (discussed below in section 7), and (3) farmers' resource constraints.

The philosophy of most packages for rainfed areas has been to replace traditional production systems with an entirely new system. An alternative approach is to begin with farmers' existing systems and introduce changes in a stepwise manner from a menu of options from which the farmer can choose (Gupta, 1991; Walker et al, 1982; Byerlee, 1994; Ryan and Subrahmanyan, 1975; Sanghi et al 1994). Such a stepwise approach would allow technologies to be introduced that are consistent with farmers' objectives and resources.

Breeding strategies for rainfed environments. The above lessons also hold for developing improved cultivars for and with farmers. A closer orientation to existing system would enable the earlier identification of priority traits for different rainfed systems, such as fodder yield and quality. In addition, there are two other unresolved challenges for breeders targeting their products to rainfed areas. The first of these is the relative emphasis to place on yield stability versus yield levels. As discussed in more detail in section 8, some evidence suggests that yield variability of HYVs is greater than for traditional varieties (Walker, 1989a). However, it is not clear if improved yield stability would significantly improve adoption and farmers' welfare, given that yield variability is not a large component of income variability (Walker and Ryan, 1990).

A closely related issue is the emphasis on broad adaptation versus narrow specificity in variety development. Since yield stability and adaptability are highly correlated (Binswanger and Barah 1980), varieties that are stable over seasons are also likely to be fairly widely adapted.

However, some observers have argued for narrow specificity of adaptation of improved cultivars and the involvement of farmers in improved cultivar selection (Maurya et al., 1988). This issue may be more relevant for rice where micro level differences in rice ecologies due to land type and water depth may be more pronounced than in dryland areas of Central and southern India. Farmers often have identified particular local varieties for these different ecologies and introduction of improved varieties will have to recognize these differences. However, for dryland crops, inter-year differences may be more pronounced than inter-field differences, suggesting the need for more widely adapted materials. In fact, the record of adoption (see Section 3 above) supports the fact that widely adapted materials are also widely adopted materials.

Research emphasis: breeding, crop management or SWM. A further unresolved issue is the extent that research for rainfed areas should emphasize cultivar improvement, better agronomic practices or soil and water management. The answer to this question is probably quite location specific. In addition, there are significant complementarities between the three technology groups. The above review has shown that improved cultivars have been widely adopted in rainfed areas with a significant impact on productivity (although less than in irrigated areas).

Some agronomic practices have also been widely adopted, especially fertilizer use. However, the role of research versus other factors such as extension, improved infrastructure, and adoption of improved varieties in the spread of fertilizer use has not been isolated. In general, research has played only a minor role in the first adoption of fertilizer (Byerlee, 1994).

Finally, adoption of SWM practices has been quite low, despite considerable investment of research resources over the past two decades. Given that improved varieties offer a proven track record of increasing productivity in many rainfed areas, there is a strong case for maintaining continued strong crop breeding programs. However, to the extent that the private sector is now able to meet the demand for improved varieties and hybrids for some crops, there may be a case for consolidation of public sector programs. While improved varieties can continue to be the lead technology in medium and higher rainfed areas, the payoff to this effort will be higher with complementary investments in improved agronomy and SWM practices.

In addition, adoption of SWM practices may be critical to preservation of the quality of the resource base (e.g., reduced soil erosion). In the most marginal areas, improved varieties have generally had little impact and are unlikely to have much impact without adoption of practices to improved moisture supply and conservation. This implies that SWM will be the lead technology in these areas, although it may be that there are many marginal areas where improved technology of any type will have little impact. Thus the emphasis will differ by ecological region and over time in the same region. However, to better inform decisions on resource allocation, there is an urgent need to increase the capacity within the research system to measure and analyze the patterns of research resource allocation between different types of research in different regions and to analyze the impact of each type of research on productivity. Very little effort in the research system is oriented to this type of analysis.

WATERSHED MANAGEMENT PROJECTS

As mentioned above, watershed management projects have become increasingly widespread and increasingly complex in recent years. As they have come to represent the principal vehicle for the transfer of rainfed agricultural technology, it is worth examining them in detail to get an understanding of their performance to date and the issues they raise. A watershed (or catchment) is a geographic area that drains to a common point, which makes it an attractive planning unit for technical efforts to conserve soil and maximize the utilization of surface and subsurface water for crop production.

A watershed is also an area that contains socioeconomic administrative and plot boundaries, lands that fall under different property regimes, and farmers whose actions may affect each others' interests. Socioeconomic boundaries, however, normally do not match biophysical ones. In watershed management projects, mechanical or vegetative structures are installed across gullies and rills and along contour lines, and areas are earmarked for particular land use based on their land use classification.

Cultivable areas are put under crops according to strict principles of contour-based cultivation. Erosion-prone, less favorable lands are put under perennial vegetation. This approach aims to optimize moisture retention and reduce soil erosion, thus maximizing productivity and minimizing land degradation. Improved moisture management increases the productivity of improved seeds and fertilizer, so conservation and productivityenhancing measures are complementary.

Excess surface runoff water is harvested in irrigation tanks while subsurface runoff recharges groundwater aquifers, so conservation measures in the upper watershed have a positive impact on productivity in the lower watershed. Reducing erosion in the upper reaches of the watershed also helps to reduce sedimentation of irrigation tanks in the lower reaches.

The watershed approach enables planners to internalize such externalities and other linkages among agricultural and related activities by accounting for all types of land uses in all locations and seasons. This systems-based approach is what distinguishes watershed management from earlier plot-based approaches to soil and water management. The World Bank-funded Pilot Project for Watershed Development in Rainfed Areas, which began in 1984, typifies the experience of the early, large scale integrated watershed projects.

Covering four watersheds averaging about 25,000 ha in four states — Madhya Pradesh, Maharashtra, Andhra Pradesh and Karnataka — these projects pursued a highly technocratic approach to soil and water management and afforestation. Trees, pasture grasses and vegetative soil conservation barriers were planted, and farmers were told to maintain them. Project managers soon realized, however, that unless inhabitants of watersheds were convinced of the benefits of the new technologies introduced under the project, they would not maintain them. The experience of this and similar watershed projects gave rise to the new calls for people's participation.

"People's participation" gradually became a buzzword for watershed management and other rural development projects in the late 1980s and early 1990s. While virtually everyone agrees that it is a good idea, however, different people define participation in different ways.

Two extremes help to characterize the experience to date with participatory watershed management. One extreme is based on the view that people will accept watershed technology once they are made aware of its benefits; this requires a mechanism for project officials to explain to watershed inhabitants what the work involves, how the various recommended practices operate, and why it is important to adopt and maintain them.

Taking people's involvement a step further, in such projects local committees are established to mobilize laborers for moving earth and planting vegetation, and to facilitate communication within the village to improve the management of common lands. The opposite extreme is based on the view that people know best how to take care of their land and simply need outside assistance to help organize them and gain access to resources, including funds and social services.

Under this approach, project officials develop mechanisms for local people to organize themselves, work collectively, and explain their priorities for external assistance. Watershed projects that emerge from such a process tend to pursue a combination of local soil and water management technologies with improved agronomic practices.

While not necessarily reflecting these two broad extremes, most government projects tend to operate closer to the first while many NGO projects resemble the second. Some of the major issues confronting these projects are as follows:

Technical vs. Socioeconomic Orientation

As mentioned above, a watershed is both a technical and socioeconomic unit. While both NGO and Government projects address both biophysical and socioeconomic relationships in the watershed, government projects devote relatively more attention to technical relationships while NGOs devote relatively more to socioeconomic relationships. This can be seen in differences in the composition of project staff and the approaches to social organization of watershed inhabitants, administrative organization, and technology choice.

Staff. Compared to NGOs, government project staff have more technical training but less experience in working closely with community organizations.

Village-level institution building

NGOs approach watershed management as one of a range of rural development activities that are chosen on the basis of villagers' priorities. Since villagers rarely consider soil and water management to be their most pressing need, watershed efforts often follow other types of projects that focus on income generation and rural health, for example. Such projects tend to focus on building community-level institutions that build organizational skills and support collective action.

These institutions, usually consisting of people of similar socioeconomic characteristics facing similar concerns, become an asset once watershed management efforts are underway. Government projects, on the other hand, also attempt to build local institutions to help implement watersheds and build support among the people to maintain them over the years. In these projects, watershed management (including treatment of private and common lands and support to livestock owners) is introduced in isolation from other activities, and village-level organizations are established solely to support the watershed project. Relatively less staff time is devoted to support the local organization, and more is devoted to technical issues. This reflects the composition of project staff mentioned above. Administrative organization. Government projects are administered at the watershed and microwatershed level, while NGO projects are more likely to be administered at the village or miniwatershed level. (A microwatershed usually is larger than a village and may contain parts of several villages, while a miniwatershed usually is smaller than a village and usually is part of only one village.)

Clearly there is a tradeoff between planning according to the optimal biophysical unit (the watershed) as opposed to the optimal socioeconomic unit (the village). Operating at the watershed level reflects the view that watershed management is primarily a technological problem, whereas planning at the village level reflects the view that watershed management is primarily a problem of social organization. It is important to stress that planning at the watershed and village levels need not be mutually exclusive, since watersheds consist of villages.

With relatively small adjustments, government projects could use the village as the planning unit and still adhere to watershed principles. Relatively minor deviations from either the socioeconomic optimum or the technical optimum probably would generate overall gains.

Technology choice. An analogous conflict takes place at the ground level in implementing project works. Technologically optimal watershed management requires that all bunds and ditches and cultivation adhere to contour lines. This approach, however, conflicts with plot boundaries, which generally are aligned to slopes but not precise contours. Contour-based watershed technologies interfere with farm boundaries and traditional cultivation practices, imposing opportunity costs on farmers and slowing adoption. Government projects adhere more strictly to contour based barriers and cultivation, while NGO projects are more willing to accept barriers and cultivation across the slope, aligned to plot boundaries.

Also, government projects follow strict guidelines regarding the dimensions and materials of bunds and other structures, whereas NGO projects are more flexible, depending on farmers' preferences and locally available materials. Because watershed management involves both physical and social processes, a compromise needs to be struck that is both reasonably technically efficient and socially workable. More information is needed on the tradeoffs involved in order to determine what is the most cost-effective approach.

The Scale of Operations

Government and NGO projects tend to operate at very different scales. Government projects cover vastly more territory, both relative to the size of the staff and in absolute terms. This makes it difficult to compare the performance of NGO and government projects. Given the small scale and higher intensity of human capital that characterize many NGO projects, better performance than government projects is to be expected. This makes it difficult to assess the effect on performance of differences in technology choice and social organization, described above.

Another difficult question regarding the comparison between NGO and government projects concerns the feasibility of "scaling up", or replicating the NGO approach to social organization on a larger scale. The same degree of human capital intensity may not be feasible in a large scale project, which in turn may constrain the degree of attention devoted to social organization. This is a question that requires further consideration, and to which several development agencies and researchers are devoting attention these days (e.g. the ODA watershed project in Karnataka).

Employment Objectives vs. Watershed Objectives

All government watershed projects and most NGO projects double as employment generation schemes. Government projects, for example, generally pay 90% of the cost of works on private lands and 100% or works on common lands. Many NGO projects offer the same subsidy as government projects, and practically all NGO projects offer a subsidy of at least 50%. The majority of costs are for labor. Given that the minimum wage paid under the project often exceeds the market agricultural wage, employment under the project is an attractive activity for laborers and many farmers. As a result, in some cases employment generation is the most important project component from villagers' perspective (Kerr and Pender, 1996a). This creates the risk that villagers will accept project activities solely to gain employment, with little interest in pursuing recommended practices or maintaining engineering works or protecting permanent vegetation. There is ample evidence that this problem occurs (Kerr, Sanghi and Sriramappa 1995). While generating employment is a worthy objective that appeals to villagers, if it interferes with watershed development objectives then perhaps the two should be separated. If so, perhaps employment could be generated more cheaply and watershed management could be promoted more effectively. This question needs examination.

Watershed Evaluations are Scarce

To date there are few comprehensive evaluation studies of integrated watershed management projects. Some evaluations show considerable impacts on adoption of some practices and on yields (see for example, the studies in the Indian Journal of Agricultural Economics, Dec., 1991, referred to below in section 6). A few studies have computed ex post cost-benefit ratios and shown a favorable ratio to the investments made. However, inadequate data limits rigorous measurement of net benefits; there are few systematic assessments of 1) the extent of adoption of watershed technologies and 2) their impact on crop yields, runoff management and soil conservation. Several reasons underlie the scarcity of information on project performance. First, detailed farm-level data on crop yields and soil and water management are expensive and time consuming to collect. Experimental data are available from watershed trials, mainly from research stations. Some experimental data were collected on trials conducted on farmers' fields but managed by project staff, so they do not reflect true farm-level conditions. Many studies estimate net benefits of government or research station watershed projects by calculating actual project costs, assuming yield impacts based on experimental data, and assuming adoption and maintenance by farmers (e.g. Singh et al 1989). This approach is not useful because too many assumptions are made in the absence of real data. On the whole, watershed projects represent a lost opportunity to collect detailed data that would offer clearer information about performance to date.

There is little evidence of sustained adoption of SWM technologies. Measuring initial technology adoption is meaningless because it is so heavily subsidized. Only revisiting project sites after the project has ended to assess the extent to which farmers continue to maintain watershed practices would give accurate information about technology acceptance. Few if any watershed projects commission evaluations of this nature. Some evidence suggests, however, that even the subsidized engineering works have often been neglected on the termination of the project (Kerr, Sanghi and Sriramappa, 1992). Rigorous benefit-cost analysis is limited to a few highly successful, highly publicized projects with a heavy infusion of technical assistance from state universities or ICAR institutes, etc. (e.g., Singh et al, 1991; Dhyani et al, 1993). A few projects, such as Sukhomajri in Haryana and Ralegaon Siddhi in Maharashtra, have received disproportionate attention. These projects, however impressive, should not be taken too casually as replicable models for other watershed projects. Both projects have certain unique characteristics not necessarily replicable elsewhere, such as vast administrative support, favorable topographic features, and single caste social structure in the case of Sukhomajri, and exceptionally charismatic leadership in Ralegaon Siddhi. It is questionable whether these successes can be replicated on a wide scale.

The few watershed evaluation studies that have been conducted raise a number of issues with regard to design and implementation of IWM projects:

- What is the role of subsidies and credit versus farmers' contribution of in kind resources, especially labor, in adoption of SWM technologies?
- How can projects increase the participation of local people in the design and management of the project?
- Can project complexity be reduced through focusing on a few components that are most relevant in the local situation?.
- What institutional mechanisms can be used to design projects which are replicable over a wider area and sustainable in the long term?

6

POVERTY AND RAINFED AGRICULTURE

Measuring poverty and relating it to the growth of rainfed agriculture is a complex topic that we treat only briefly here. We refer to existing literature to review various aspects of rural poverty, and we discuss poverty alleviation measures and their strengths and weaknesses. The literature on poverty in India is large, so we cannot do justice to it. Table 6.1 shows changes in the percentage of people falling below the poverty line in different states between the years 1973-74 and 1987-88.

Unfortunately this table does not give us any indication of the distribution of poverty between rainfed and irrigated areas or across different rainfed types, but it does show some important trends. In particular, the percentage of rural people who are considered poor by the poverty line indicator has fallen steadily in every state listed.7 This constitutes part of the evidence for what Singh (1990) refers to as the beginning of the "great ascent" out of poverty in rural South Asia.

Rainfed regions are characterized by higher poverty than irrigated regions. Table 6.2 shows clearly, for the year 1973, that the percentage of the population below the poverty line falls steadily as irrigated area rises. The district database does not include information on the number of people below the poverty line, so a rigorous analysis of the relationship between irrigated area and percentage of people in poverty is not attempted here. State level data are available, however, and can give us some indication.

Table 6.1 Changes in percentage of rural people falling below poverty line in major states, 1978-1988

State	Percentage of Rural People Falling Below Official Poverty Line		
	1977-78	1982-83	1987-88
Andhra Pradesh	48	38	27
Bihar	63	63	64
Gujarat	46	42	30
Haryana	34	28	21
Karnataka	55	48	36
Kerala	59	52	39
Madhya Pradesh	63	63	49
Maharashtra	58	64	45
Orissa	67	72	68
Punjab	28	16	13
Rajasthan	45	36	34
Tamil Nadu	57	58	54
Uttar Pradesh	57	48	46
West Bengal	73	68	63

Source: Tables 4.2-4.5, Report of the Expert Group on Estimation of Proportion and Number of Poor. Planning Commission, 1993.

Table 6.3 shows the relationship between the number of people below the poverty line and the percentage of rainfed area at the state level for the years 1983 and 1987. The correlation coefficient is 0.45 for 1987, suggesting a fairly strong relationship. The state level data, however, are too aggregated to be very precise, and they do not distinguish between rural and urban poverty. Also, in this approach tiny states like Tripura and Sikkim receive as much weight as large states like Uttar Pradesh and Bihar. Therefore we should not draw strong conclusions from this table.

Table 6.2 Percentage of population below the poverty line in relation to the percentage of area irrigated (1973)

Gross Irrigated Area as Percent of the Gross Cropped Area in the Triennium Ending 1973	Number of Regions	Percentage of Population Below the Poverty Line
Below 10 percent	16	68.75
10 - 20 percent	13	53.70
20 - 30 percent	10	45.62
35 -50 percent	8	48.39
Above 50 percent	7	26.46

Source: Rao et. al. (1988)

As mentioned above, rural wages are an imperfect indicator of poverty. Repetto (1994) examined changes in agricultural employment associated with growth in agricultural output in the 1970s and 1980s. He found that on-farm employment actually fell in Haryana and Punjab, where agricultural output grew the fastest. This information alone has little implication for rural poverty, however, considering the fact that these two states have by far the lowest levels of rural poverty in the country, in part because the green revolution also stimulated strong growth in the nonfarm economy. Still, Repetto points out that over the years, agricultural growth has had a less than commensurate effect on employment growth. He cites Basu and Kashyap (1992) as finding that the employment elasticity of agricultural growth declined continuously during the 1970s and 1980s.

Table 6.3 Poverty and percentage of rainfed areas by state, 1983 and 1987

	1983		1987	
State	% People Below Poverty	% Rainfed Area Line	% People Below Poverty	% Rainfed Area Line
Andhra Pradesh	29.88	63.40	27.20	62.70
Assam	40.86	84.60	36.84	84.30
Bihar	62.51	63.30	53.37	63.30
Gujarat	33.27	74.50	32.33	75.40
Haryana	21.24	36.40	16.63	30.90
Himachal Pradesh	16.39	82.90	15.46	82.90
Jammu & Kashmir	24.10	59.50	23.20	60.60
Karnataka	38.47	82.00	38.14	81.00
Kerala	40.91	85.30	32.08	85.20
Madya Pradesh	50.13	86.10	43.40	84.40
Maharashtra	43.54	86.90	40.10	87.60
Manipur	38.08	59.70	32.93	60.30
Meghalaya	39.46	76.40	34.60	75.20
Nagaland	39.75	71.70	34.85	71.20
Orissa	65.32	76.90	55.61	77.50
Punjab	16.29	9.50	12.70	8.70
Rajasthan	35.02	77.80	34.60	75.30
Sikkim	39.62	87.20	34.67	88.10
Tamil Nadu	52.38	50.50	45.13	56.30
Tripura	40.79	89.30	36.84	89.20
Uttar Pradesh	47.19	51.60	41.99	49.00
West Bengal	54.72	73.90	43.99	76.70
All India	44.76	69.30	39.34	68.60
Correlation coefficient		0.35		0.45

Source: Planning Commission 1993 and Area and Production of Principal Crops in India, various issues

Each percent increase in agricultural output yielded an increase in agricultural employment of 0.7% in the 1970s but only 0.3% in the 1980s.

There has been a long debate about whether the productivity gains of the Green Revolution resulted in welfare improvements for the rural poor. Ravallion and Datt (1994) cited two studies, one that argued that "trickle down" of benefits to the poor has been a reality, and another that argued that growth has led to increased inequality. These two studies addressed roughly the same period of time, so their conflicting conclusions are surprising.

More recent empirical work has suggested that agricultural growth has in fact succeeded in reducing poverty levels. Singh (1990) argued that all income groups gained from the green revolution, but said it is "widely accepted" that wealthier groups benefitted more than less wealthy groups.

Hazell and Ramasamy (1991), on the other hand, show evidence that the large income gains associated with the green revolution between 1973-74 and 1983-84 in North Arcot, Tamil Nadu, did benefit landless and small farm households by at least the same proportion as for large farm households. They found that a 4% decrease in total agricultural employment was more than offset by large wage increases and increased off-farm employment opportunities.

Also, the lower on-farm employment figures reflected a drop in labor market participation by members of households farming more than one hectare of land rather than the landless or holders of even smaller plots. Larger farms also gained from the green revolution, but less so due to larger cost increases they faced due to rising labor and fertilizer costs. The conflicting findings of Singh, on the one hand, and Hazell and Ramasamy on the other may reflect differences in the rate of adoption of green revolution technology. Hazell and Ramasamy found that larger farmers adopted much more rapidly than small farmers, but after ten years the use of green revolution technologies was scale neutral.

They also showed that smaller farmers were able to retain their land holdings despite their initially inferior position, enabling them to remain in a position to make strong gains after they eventually adopted. Also, Hazell and Ramasamy only studied one small region.

Hazell and Ramasamy also found that every additional rupee of value added in agriculture stimulated growth in value added in the nonagricultural sector equivalent to Rs 0.80. This large multiplier effect has important implications for the role of agriculture in stimulating widespread economic development.

It also helps to offset concerns that may be generated by the reports of the small impact of agricultural growth on agricultural employment, cited above. Ravallion and Datt (1994), in a study of household data from 1951 to 1991, drew conclusions similar to those of Hazell and Ramasamy. They found that rural economic growth spread gains evenly and so brought ample gains to the rural poor; rural growth even helped reduce urban poverty to a certain extent. A related debate addresses the question of whether urban or rural economic growth has the greater impact on poverty alleviation.

Ravallion and Datt (1994) found that urban economic growth was associated with unequal distribution of benefits, so that its impact on urban poverty alleviation was small and the impact on rural poverty was insignificant. This contrasts sharply with their findings above regarding the favorable impact of rural economic growth on poverty alleviation. Ravallion and Datt concluded that both continued rural economic growth — including growth in rainfed agriculture — and a more equitable process of urban growth will be important to poverty alleviation in the future.

SPECIAL PROGRAMS FOR POVERTY ALLEVIATION

The relatively favorable performance of better endowed rainfed regions was documented in sections 3 and 4. Continued increases in productivity in such regions can be expected to help reduce poverty, particularly in combination with efforts to improve infrastructure and institutions critical for delivery of services. Likewise, widespread poverty is likely to remain for a long time in poorly endowed areas less likely to enjoy high agricultural growth and rising wages. For these areas even more than others, migration and development of the nonagricultural sector must provide the solutions. Singh (1990), von Braun (1995) and others, however, comment that while economic growth and diversification may be the solution to long term poverty alleviation, the process takes too long to solve today's poverty problems. Growth on the scale required cannot happen overnight, and it takes time for it to have a significant impact on poverty.

This is the rationale for the introduction of numerous special antipoverty programs introduced in India beginning in the 1970s. The programs are intended to alleviate poverty in the short term while seeking to stimulate economic growth that will reduce poverty in the long term. Of the many central government schemes initiated in India, the largest are the Integrated Rural Development Program (IRDP), introduced in 1978-79, and the National Rural Employment Program (NREP), which began in 1980. The IRDP mainly provided subsidies and credit to purchase livestock and milch animals, outlaying Rs 10 billion and covering about 3 million families per year in the mid 1980s. This program, however, suffered from difficulties in targeting benefits to the poor and encouraging repayment of loans. Despite its large coverage, it paled in comparison with the needs of the estimated 260 million rural people.

The NREP focused mainly on employment, generating 350 million additional days of employment per year through Food for Work programs that executed a wide variety of public works projects, including roads, drinking water projects, small scale irrigation works, soil and water conservation and afforestation (Dantwala, 1986, cited in Singh (1990).

Singh and Dantwala argue that employment programs have a greater capacity to help the poorest people because they can be self-targeted, with less spillover than other measures such as food aid, and they can also create durable infrastructure that leads to development. Von Braun (1995) presents a conceptual model and detailed evidence of the positive contribution that employment programs can make to poverty alleviation.

He shows why employment generation can be a superior alternative to other measures such as food subsidies, which benefit poor people but are expensive, distort food prices, and suffer significant leakage to nonpoor households due to targeting difficulties (Pinstrup-Anderson, 1988; Repetto, 1994; Parthasarathy, 1995).

Employment programs, on the other hand, can be self-targeted toward poor people whose only productive asset is their labor. Employment programs can, in principle, be self-financed though taxation that pays not only for employment but productive assets and improved infrastructure that employment generates. In India, employment programs have been used for constructing canal irrigation schemes, building roads and drinking water wells and buildings. Eventually some concern arose that the opportunities for productive, labor intensive infrastructure development were dwindling

(Thomas Walker, pers comm.). The amount of unutilized canal irrigation potential dwindled, virtually every village had a school house and a health center, and road work was confined mainly to maintenance, not construction of new roads.

With the increased focus on developing rainfed agriculture over the last decade, there has been an increased focus on using employment to support rainfed agriculture. Watershed development, including soil conservation bunds and ditches, check dams and tree planting, appeared to be a useful way to support the development of rainfed agriculture in distressed areas while also providing short term employment (Hanumantha Rao, 1992).

The focus on asset creation to support rainfed agriculture is attractive, but there is insufficient information regarding its effectiveness to date. Jackson (1992), in a worldwide study that included India, argued that the asset-creation impact of employment programs was exaggerated due to the low quality of work they produce.

Kerr et al (1994) argue that in India, watershed development efforts linked to public employment programs are unlikely to have a lasting impact on rainfed agriculture under current program design. They present numerous field observations suggesting that soil and water conservation programs based on employment programs may create illusory gains.

This is because in order to obtain employment at the minimum wage, which in some cases exceeds the market wage, participants are encouraged to install mechanical structures that they may dismantle immediately afterward, or to plant trees that they will not protect. For this reason, the link between public employment and rainfed agricultural development may be more complex than first thought. In fact, in several countries, particularly in Africa and the

Caribbean, employment subsidies have been separated from soil and water conservation programs because the objectives of employment and natural resource management appeared to be in conflict (Kerr et al, 1994).

The point is that long run asset-generation effects of employment programs should not be taken for granted, particularly as related to rainfed agriculture. The papers in Von Braun (1995), for example, examine in great detail the benefits of employment but largely assume the durability of assets created. They devote more attention to whether such assets will result in equitable income distribution than the amount of such assets they will actually generate. Further research and field experimentation are needed to understand the nature of incentives they create and the steps needed to ensure that they are consistent with long term asset creation as well as short term poverty alleviation. Kerr et al (1994) suggest some ideas tried on the ground in India and Africa to achieve that objective.

Still, it may prove to be that some kinds of assets, like roads, wells and buildings can be created more effectively through employment programs than others, like trees and soil conservation bunds. Other problems also remain with employment programs. Jackson (1992), for example, argues that such programs often are associated with high leakage. Unpublished ethnographic data from ICRISAT (1993, 1995) also cites villagers' complaints of leakages and favoritism in distributing employment benefits, but there is no evidence on the scale of the problem.

On the other hand, such government programs will always be associated with some leakage, but that does not necessarily mean that they should be eliminated. In any case, there is room for improvement in making employment programs more cost-effective. Finally, one important aspect of employment generation programs is their capacity to be self-targeting, with minimal distortion to the rural economy. These objectives can be achieved if wages are kept slightly below the going market wage, and if they are concentrated in the slack season, when employment demand is low. If wages are low, only those people who really need them will participate.

If they are high — above the market wage, in particular — then employment will be rationed and some people who need it will not be able to get it. Ravallion, Datt and Chaudhuri (1990), for example, found that raising wages may cause a greater proportion of the poorest people who seek work to be excluded. This is so because as program benefits become more attractive and have to be rationed, less powerful people will have a more difficult time bidding for them.

Low wages also ensure that small farmers who rely on hired labor during peak periods do not suffer from having to pay high wages due to labor scarcity created artificially through public employment programs. Legislation in India requiring public employment programs to pay the national minimum daily wage of Rs 22, though well-intentioned, may cause employment programs to be less effective in reaching the poorest people, make them financially unaffordable, and damage the interests of small farmers by raising wages above the market wages that prevail in some areas.

GRASSROOTS ANTIPOVERTY INITIATIVES THAT SUPPORT RAINFED AGRICULTURE

Over the years, Indian NGOs have experimented with a wide variety of grassroots approaches to rural development. Some of these, honed by a relatively small number of high quality NGOs, appear to have promise for poverty alleviation and increased productivity even in marginal areas. The wide variety of approaches to community development defy easy categorization, but to be brief, we outline a stylized example of the experience to date.

Many NGOs find that informal savings cooperative groups are high on the list of villagers' priorities for development assistance (Fernandez, 1991; G. Sriramappa, Oxfam, pers comm.).

Savings groups are established loosely, with members selecting the composition of their own groups, and many groups, typically of homogeneous membership, often operate simultaneously in the same village.

Experience suggests that participation in these groups helps people generate capital, develop organizational skills, build villagers' confidence to work collectively, to seriously consider new investment opportunities, and to act where they never bothered to previously (James Mascarenhas, OUTREACH, personal communication; Parthasarathy, 1994). Active local groups can stimulate psychological incentives that previously were stifled by cultural or political constraints.

Mascarenhas, et al (1991) and Fernandez (1993) discuss links between promotion of such groups and the development of rainfed agriculture through watershed management.

First, participation in credit groups can help villagers save funds to invest in rainfed agriculture. One common experience in MYRADA's and Oxfam's thrift group projects is that even without any financial assistance, within a year or two participants find that their greatest challenge is to figure out how to spend their savings.

Second, developing and strengthening cooperative groups prior to tackling watershed development helps villagers build organizational and conflict resolution skills that are important in watershed management.

Third, the groups can serve as a focal point for spreading awareness about the benefits of watershed development. In MYRADA watershed programs, existing cooperative groups developed to solve an unrelated set of problems become a focal point for development of rainfed agriculture.

Small group credit generation programs have multiplied rapidly in south India. MYRADA's and Oxfam's thrift projects, for example, have tens of thousands of participants, and they are spreading rapidly. Information about these projects is limited, and their potential to promote rainfed agriculture through indirect means is not well understood. They are worthy of further attention, both to understand the role they can play in alleviating poverty and promoting rainfed agriculture, and to understand how to extend such an approach to a state or national scale.

7

NATURAL RESOURCE DEGRADATION AND RAINFED AGRICULTURE

Natural resource degradation in rural areas is a controversial topic. There is a wide range of opinions regarding its causes, extent, consequences, and even its definition. Narrowly defined, degradation of a natural resource can refer to a permanent, irreversible loss of its productive capacity relative to its natural state. In a broader, socioeconomic perspective, on the other hand, often human use of natural resources may be taken for granted, so the natural state is not particularly relevant.

Productive capacity must be defined in terms of a given use; exploitation of a natural resource may lead to a situation in which it becomes unsuitable in one type of use yet still productive in another. Equally importantly, in some cases losses in productive capacity at one point in time may be reversed later, because human activity can also improve the productive capacity of natural resources.

In discussing degradation, therefore, it is important to be explicit about what we are talking about. We follow Scherr et al (1995), who suggest that for policy purposes, a dynamic view of natural resource degradation and improvement is most useful because it allows us to examine the question of how human exploitation affects natural resource productivity under particular uses.

There are many types of natural resource degradation, such as deforestation, soil erosion and other types of soil degradation, loss or pollution of surface water or groundwater supplies, and loss of resistance to pests and diseases. All of these have some relationship to rainfed agriculture. Data on natural resource degradation are relatively sparse, and information on the implications for productivity are still more rare. We draw on a small amount of published data in this section to present some evidence about the extent and implications of various forms of natural resource degradation. Then we discuss some determinants of people's actions that lead to degradation or improvement in natural resource productivity.

SOIL DEGRADATION

Soil is the natural resource whose degradation causes the most widespread concern about rainfed agriculture in India. Soil degradation comes in several forms, including erosion by wind or water, and chemical deterioration such as loss of nutrients or salinization. Sehgal and Abrol (1994) compile the available information on soil degradation and conclude that erosion by water is the most widespread form of land degradation in India. Table 7.1 shows ICAR's official estimate of the distribution of lands affected by different types of degradation in the country. Water erosion is distributed throughout the country; its most severe form is found in parts of Uttar Pradesh, Madhya Pradesh, Rajasthan and Gujarat that are prone to wide, deep ravines that disfigure the land and make agriculture impossible. In other areas, erosion gradually removes topsoil and reduces yields, but it does not cause much land to be taken out of production in the short term. Wind erosion in India is confined to desert regions of Rajasthan and Gujarat, but it is not a significant problem elsewhere. Salinity is a serious problem affecting mainly irrigated areas; it causes severe productivity losses on high potential lands; some lands are even taken out of production. Lands with other nutrient losses are surprisingly small in area given the large area of drylands with soils that are naturally deficient in nutrients and receive only small applications of fertilizer and organic matter.

Table 7.1 ICAR's estimate of area of land subjected to various forms of degradation (million hectares)

Status	Land subject to degradation (*including* land slightly affected)		Land subject to degradation (*excluding* land slightly affected)	
	Area	Percent of total	Area	Percent of total
Water Erosion	148	45	122	37
Wind Erosion	14	4	13	4
Chemical Deterioration	14	4	11	3
Waterlogging	12	4	5	2
Not fit for Agriculture	18	6	18	6
Total Area Degraded	187	57	169	46
Unaffected Area	123	37	160	49
Total Area	329	100	329	100

Source: Sehgal and Abrol, 1994.

Note: Numbers may not add up due to rounding errors.

Sehgal and Abrol's estimates of area under erosion exceed most of those estimated elsewhere. Table 7.2 shows the figures derived from other studies. The various studies address slightly different questions and use different methods to arrive at estimated figures, so their comparability is limited. We use the Sehgal and Abrol study because it is based on the most systematic assessment, following the guidelines in the Global Assessment of Soil Degradation (GLASOD) (Oldeman, 1988). Sehgal and Abrol rely on several sources of information, including a generalized soil map of India, remote sensing data from selected areas, and published information on forestry and different soil degradation problems.

Estimated Rates of Soil Erosion and its Consequences

Estimates of rates of erosion and its consequences vary. Also, the impact on productivity of a given rate of soil loss depends greatly on the soil type, its original depth, and the time period over which the rate of erosion is sustained. Aggregate figures on soil loss per hectare can be misleading, because often they are estimated on the basis of soil loss measurements on individual experimental plots. Often under this approach, all soil that erodes in a given area is assumed to disappear, to be lost permanently. Observations on farmers' fields in India, however, reveal that sometimes erosion can be a loss to one farmer but a gain to another on whose land the eroded soil is deposited. Many farmers actually encourage erosion on one part of a plot in order to concentrate soil where it can contribute to greater overall productivity (Kerr and Sanghi, 1992). What this means is that a given estimated rate of soil loss for a given region probably is accurate on some plots or parts of some plots, but other areas are probably losing a smaller amount and some are probably even gaining more than they are losing. With this caveat in mind, we present briefly some estimated rates of soil loss for India. Singh et al (1992) produced an iso-erosion map of India that delineates areas by their rate of soil loss per ha. On the high end, ravine areas (mentioned above) are estimated to lose over 40 tons/ha per year, and the Shivalik Hills at the foot of the Himalayas are thought to lose over 80 tons/ha/yr. Rates in the Indo-Gangetic Plains are estimated at uniformly less than 10 tons/ha/yr, and those in peninsular India, excluding mountainous areas, range from less than 5 to over 30 tons/ha/yr.

Table 7.2 Estimates of degraded area in India

Study	Degraded Area (million ha)	Comments
FAO/RAPA (1992)	172	Includes 127 m ha subject to erosion, 29 m ha fertility decline, 9 m ha waterlogging, 7 m ha salinization alkalinization.
Dregne and Chou (1992)	102	This figure only refers to dry areas (but not deserts) totaling 163 m ha according to the authors. Their figure includes 60 m ha rainfed lands, 8 m ha irrigated lands, and 34 m ha rangelands. Forests are excluded. As a percentage of the area studied, this estimate is relatively high.
Bentley (1984)	115	Bentley defined wastelands as land currently producing less than 20% of biological potential. His figure includes 15 m ha of marginal agricultural lands and recently deforested lands.
Bhumbla and Khare (1984)	93	If nonforest wastelands are included, this figure becomes 129 m ha, the figure accepted by the National Wastelands Development Board.
Gadgil et al (1982)	88	This estimate excludes cultivated lands. Degraded lands distributed

		as follows: pasture lands 12 m ha, degraded forests 36 m ha, culturable waste 17 m ha, fallows 23 m ha.
Vohra (1985)	103	Distribution is as follows: 30 m ha forest land, 33 m ha uncultivated land, and 40 m ha crop land.
World Bank (1988)	115-130	Includes 32-40 m ha of degraded land; the rest is similar to Gadgil et al (1982).
Chambers, Saxena and Shah (1989)	109	Refers to lands producing substantially below potential. Distribution is 38 m ha cultivated lands, 2 m ha strips and boundaries, 36 ma ha degraded forest land, and 33 m ha uncultivated degraded lands.

Dhruvanarayana and Ram Babu (1983) attempted to aggregate soil loss throughout India. They estimated an average nationwide soil loss rate of about 16 tons/ha/yr, with about 29 percent of eroded soil being permanently lost to the sea. They also estimated that about 9 percent of the total soil lost nationwide was deposited into major reservoirs, reducing their capacity by 1-2% annually. The remaining quantity of soil lost simply moved from one place to another.

El-Swaify et al (1982) suggests that 11 tons/ha/year is an acceptable rate of erosion on most soils because it is the rate of natural formation of new soil. Beyond 12 tons/ha, erosion will cause soils to become continually shallower. Unfortunately, the consequences of soil erosion are no better understood than the rates. According to Dregne, quoted by Crosson (1994), there is an "abysmal lack of knowledge" about the productivity implications of soil erosion. Measuring erosion and its productivity effects is expensive and time consuming, especially in semi-arid areas where weather changes from year to year, so that data must be collected over several years in order to understand biophysical relationships correctly.

This situation has led to increased interest in crop simulation models which, once they are validated under a wide range of conditions, can provide information on erosion rates and productivity implications quite quickly.

Simulation models estimate the likely amount of runoff under different agroclimatic conditions, such as rainfall, slope, length of slope, soil type and soil cover, and the likely amount of erosion associated with that runoff under those conditions. The models incorporate probable weather patterns based on decades of daily rainfall data.

Then they use crop growth models to estimate the impact on productivity. Littleboy et al (1996) used the PERFECT model to relate yields to soil depth and simulated soil loss for given soil conditions. They calibrated the model to conditions on red soils at ICRISAT, near Hyderabad, and found relatively little short term yield decline, but a permanent, catastrophic loss in yields after 40-90 years of soil erosion.

The time horizon, of course, depends on erosion rates associated with different soil management practices. Their findings showed that alfisols with slope greater than 5% should not be cropped. Other simulation models of soil loss have not been calibrated to agroclimatic conditions in India, but there is increasing activity in this field, and new, more accurate information will probably be available in coming years.

Sehgal and Abrol (1994) present estimates of the severity of erosion in India. They cite experimental data on erosion-productivity relationships based on three-year experiments on vertisols in Nagpur, Maharashtra, that shows very small losses if erosion is kept to less than 10 tons/ha/year, but very high losses under higher rates of erosion. They find that erosion at the rate of 10-20 tons/ha/year causes yields to fall by 7.8-34.3% per year, and erosion at the rate of 20/40 tons/ha/year, causes losses of 58-68% per year.

These findings suggest that for most crops, erosion at rates of more than 10 tons per year would lead to practically zero yield in less than two decades (table 7.3). Such rapid losses in productivity are not without precedent. Alison (1973), for example, says that millions of acres throughout the United States suffered such heavy rates of erosion that cultivation became unprofitable until abundant use of fertilizer made it possible to use them again.

The restoration of such soils raises an important point regarding the distinction between reversible and irreversible effects of erosion. Reduced soil depth is obviously an irreversible effect of erosion, unless a plot actually receives soil deposits resulting from erosion further up the slope. A sustained reduction in soil depth will eventually make cultivation impossible, as mentioned above in the discussion of the findings of the PERFECT model.

Other losses associated with erosion, on the other hand, are reversible albeit at a potentially high cost.

It is well-known that as soil erodes, disproportionately large amounts of organic matter tend to be lost. According to Allison (1973), eroding soil can contain up to five times as much organic matter as soil left behind.

Experiments at ICRISAT found that on plots that received applications of farm yard manure, it was found to be heavily concentrated in eroded material (QDPI/ICRISAT, 1991, pg 6). Water holding capacity also drops with erosion; this takes different forms in the short and long term. In the long term, water holding capacity is permanently reduced because shallower soil contains less space

Table 7.3 Loss in crop productivity (yield) at different degrees of erosion, slope and soil depth

Class Range	Loss in Productivity/Actual Yield Q/ha and (percent loss)				
	Sorghum	Cotton	Pigeonpea	Groundnut	Soybean
			Soil Parameter: Erosion		
e1: Slight	0.8	—	—	1.0	—
(5-10t ha yr)	(2.5)	(—)	(—)	(5.8)	(—)
e2: Moderate	5.9	1.4	4.1	5.9	6.0
(10-20t ha^{-1} yr^{-1})	(18.5)	(7.8)	(21.5)	(34.3)	(29.8)
e3: Strong	21.7	11.0	11.1	11.7	12.7
(20-40t ha-1yr-1)	(68.0)	(61.1)	(58.1)	(68.0)	(62.9)
			Soil Parameter: Depth		
Deep	—	—	—	2.1	2.7
(>100 cm)	(—)	(—)	(—)	(12.2)	(13.4)
Medium	14.1	8.4	5.0	0.6	3.0
(50-100 cm)	(44.2)	(46.6)	(26.2)	(3.5)	(14.9)
Shallow	23.4	6.7	11.2	9.2	12.7
(<50 cm)	(73.4)	(37.2)	(58.6)	(53.5)	(62.9)
Model Yield*	31.9	18.0	19.1	17.2	20.2

* Model yield with soil-site suitability at optimum (suitable) level

Farmers 8 were asked to estimate changes in yield on a given plot over several years under two cases, one in which it is subject to erosion, and the other after conservation measures are adopted.

They also indicated the amount of fertilizer and farm yard manure they applied to their fields and the amount of these inputs they expected would be lost to erosion. Regressions on the basis of these data yielded the perceived erosion-yield relationship, and also the portion of yield resulting from loss of applied nutrients compared to other factors. Unfortunately, farmers were not asked to about the effect on yields of lost moisture lost due to store water. In the short term, erosion is correlated with high runoff, indicating that when erosion is taking place, the soil captures less moisture to supply to crops. When conservation is introduced and erosion ends, however, moisture retention can be increased and organic matter content can be restored.

This distinction between short- and long term impacts of erosion is important to consider when trying to assess the aggregate impact of erosion on productivity. Sehgal and Abrol (1994), for example, combine estimates of the extent of erosion in India with those of its severity (table 7.4), and their findings are alarming indeed.

They divide severity of degradation into four categories: low, medium, high and very high. Low severity indicates that yield losses resulting from degradation are less than 15%; medium means losses of 15%- 33%, high means that 33-67% of productivity is lost, so that cultivation is uneconomical and other uses, like agroforestry, provide the only hope; finally, very high indicates that soil degradation is so great as to make the soil unusable. Given these categories, it is difficult to believe the finding that over 142 million ha suffer from "high" or "very high" severity of degradation out of a total cultivated area of 187.7 million ha. This would imply that over 75% of the land cannot be cultivated economically, a conclusion not entirely consistent with the observation that farmers continue to cultivate most of that land.

Continued cultivation is also inconsistent with the finding in table 7.3 that even moderate erosion would bring yields down to practically zero in less than a decade on all plots subject to moderate erosion. Experimental findings of high yield losses due to erosion are not greatly different from rates estimated by farmers. Farmers interviewed in three villages in Maharashtra and Andhra Pradesh estimated that yields would fall on eroding fields by an average of 5-10% per year, depending on the village (Kerr and Pender 1996b).8 One of the villages, to runoff.

Kanzara, has agroclimatic conditions similar to those in Nagpur, where the experimental data reported in table 7.3 were generated. The farmers' estimates are extremely rough, or course. Nevertheless, the reported yield loss rates make an interesting comparison to the experimental findings. Farmers' yield loss rates appear to be on the low side compared to the experimental data. Farmers' perceptions of erosion-yield relationships are consistent with the suggestion that moisture retention and organic matter content can be restored after conservation measures are introduced. Their responses suggest that farmers perceive that up to 25% of the yield decline results from the runoff of farm yard manure and fertilizer they applied to their fields. Farmers estimated that yields will rise by 3% to 14% per year when soil conservation practices are put in place (Kerr and Pender, 1996b). Although these figures should be considered as no more than rough indicators of farmers' perceptions, it is clear that they do not perceive productivity losses from erosion to be entirely irreversible. This makes sense in light of the discussion above.

To summarize this section, evidence of declining yields associated with soil erosion should be taken seriously, especially given predictions that irreversible losses due to reduced soil depth will make cultivation impossible on many soils in less than a century. In addition, experiments that show short term losses of 10%-50% resulting from erosion also imply a serious reduction in current production levels. On the other hand, clearly there is a danger in estimating all-India erosion rates by extrapolating from findings on small-plot experiments. Likewise, it is important not to assume that evidence of short term yield decline represents a permanent, monotonic relationship. More information is needed about spatial and temporal variation in erosion rates under different biophysical conditions and management practices.

Costs and Benefits of Soil Conservation Investments

Data on the costs and benefits of soil conservation benefits are scarce. Some studies have tried to estimate costs and benefits on the basis of experimental data of improved practices. Even if their erosion-productivity data are accurate, their estimates of costs may be inaccurate because, although they can capture cash costs of conservation investments, they omit opportunity costs associated with changing other aspects of the farming system in order to accommodate the improved conservation practice (Kerr and Sanghi, 1992). Other studies use actual cash cost data from watershed projects and estimate the benefits on the basis of either experimental data or observed field data for a year or two, and the assumption of adoption and continued maintenance. These studies, however, are misleading because there is little or no evidence of sustained adoption of recommended practices introduced under watershed projects. In some cases farmers may selectively retain certain components of watershed technologies (Joshi, 1995), but data are not available on the net returns to using the components in isolation. With this caveat in mind, we present some figures on the net benefits of improved soil conservation from the Indian literature. The literature on costs and benefits of soil conservation investments in India is large, even if most studies are not very comprehensive, and we lack the resources to do justice to it here. A thorough literature review of watershed and soil and water conservation literature from India will be conducted as part of the ICARWorld Bank-IFPRI-ICRISAT rainfed agriculture study. The Indian Journal of Agricultural Economics (1991) contains summaries of numerous evaluations of watershed management projects. These projects include far more than just soil and water conservation; for example, they introduce new seeds and other inputs, they try to increase grass and forest cover and, in some cases, they even develop irrigation. In addition, the summary reports use a wide variety of indicators of net project benefits, so direct comparison is not feasible. Rather than recite the exact findings of the many studies listed in the volume, we simply list a few general points.

- Many studies list cost-benefit ratios. In all cases they are greater than one, and in some cases they range up to over 2.
- Two studies calculate internal rates of return. In one study the various project components have rates of return ranging from 8-12%, including 10% for the soil conservation component. In the other study the range is 12-15%, with 12% for the soil conservation component.
- Many studies cite increased yields of various crops; they range from 10% to 100%. Most of these studies do not provide information about net returns associated with yield increases.
- Other studies cite qualitative improvements such as increased cropping intensity, rising water tables, increased irrigated area, higher input use, and higher employment.

It is important to note that many of the studies compare performance in two years, one before the project and another during the project. They attribute the improved performance in the latter year to the watershed project. This method is flawed, however, because any effect of a watershed project may be dwarfed by the effects of large swings in weather.

The studies that pursued this approach did not address this potential problem. While many studies have evaluated the net benefits of watershed projects, they have not followed a common methodology, and many of them present questionable results. Clearly, further work is needed to take a more systematic approach to comparing watershed management experiences under different circumstances.

Farmers' Adoption of Soil Conservation Practices

Little work has been done on the determinants of farmers' decisions regarding soil and water conservation investment. Some studies address these questions in the context of special soil conservation or watershed management projects but, as mentioned in section 4 the relevance of this approach is questionable, because programs are so highly subsidized that initial adoption is meaningless. A more useful approach would be to return to project sites after they have been completed in order to assess farmers' continued use of technologies and practices introduced under a project. Kerr and Sanghi (1992) and Pender and Kerr (1996) study the circumstances under which farmers invest in soil conservation. These studies focus on areas that are outside of watershed project areas, so farmers' investments are unsubsidized. As a result, the conservation practices in question are indigenous ones, not those introduced under watershed projects. The differences in the two types of technology were discussed in the section on agricultural technology.

Some of the findings of Kerr and Sanghi (1992) regarding technology design are discussed above in section 5. They presented several hypotheses on the basis of field observations and group interviews; we mention a few of them here. They suggested, for example, that farmers invest in soil conservation at least as much for short term productivity as long term conservation objectives, because soil conservation measures also conserve moisture and organic matter. As a result, farmers are more likely to invest in conservation on their most productive plots, even if they are not the most prone to long term erosioninduced productivity losses. Kerr and Sanghi also hypothesized that farmers under tenancy are unlikely to invest in soil conservation, because state laws in their study area (several south Indian states) discourage tenancy, effectively limiting tenancy contracts to no more than two seasons, and usually not more than one. As a result, tenants will not realize the long term benefits of conservation investments. Another hypothesis is that farmers are less likely to invest in soil conservation on dryland the more income they derive from alternative sources, including irrigated land. The rationale for this hypothesis is that farmers with alternative income sources will be less dependent on rainfed agriculture and thus less concerned about its long term productivity. On the other hand, if credit markets are imperfect, farmers with alternative sources of income might invest more because they are less constrained financially. A related hypothesis is that farmers with more family labor available for agricultural work will invest more in soil conservation, because the marginal product of their labor in other uses may be lower. This argument is based on the view that in many cases soil conservation is an activity with relatively low returns, so it is worth devoting a few hours of work here and there, but not necessarily worth hiring workers for a full day at the market wage. This argument would imply an imperfection in labor markets.

Pender and Kerr (1996) test these and other hypotheses about soil and water conservation investments in the same study area. Their preliminary econometric findings support the hypotheses that plots under tenancy receive less investment than owner-operated plots, and that farmers tend to invest in their best plots first. They found mixed evidence about the effects of family labor and alternate income sources, depending on the village. These differences can be explained in part by the diverse agroclimatic conditions across villages, which may change the benefits of soil and water conservation investments. This research is still in progress, and alternate specifications of the model may yield additional information or clarify some of the ambiguities in the preliminary results.

Waterlogging, Salinization and Alkalinization of Irrigated Lands

This is potentially a very serious problem for Indian agriculture because it affects the most fertile lands in the breadbasket areas of Punjab, Haryana, and northwestern Uttar Pradesh. Estimates of salt-affected lands also vary greatly, ranging from about 3 million ha (Joshi and Singh 1991) to about 25 million ha (Bowonder and Ravi 1984). Joshi and Jha (1991) found that after 10 years of salt accumulation, rice yields declined by 61% and wheat yields by 68%. Some soils in rainfed areas also suffer from salinization, but the area affected is smaller and the absolute yield decline is smaller. But salinity and alkalinity on irrigated lands affect rainfed agriculture indirectly, because it increases the need for production on rainfed lands to augment production on irrigated lands. In many places salt-affected lands are taken out of production completely, but this need not be the case. The Central Soil Salinity Research Institute has developed salt-tolerant HYVs of rice and wheat, for example. Also, large amendments of gypsum and other chemicals and minerals can remove salinity, though eventually it may return. In the longer term, better drainage is needed in order to reduce the water table in irrigated areas. In many cases this requires significant up-front investment, and also collective action among area farmers to ensure that drainage is managed properly across several farms. Joshi (1996) discussed the conditions under which farmers will invest in soil amendments to neutralize soil salinity by comparing the cases of Punjab, Haryana and Uttar Pradesh. He found that the extent of investment depends greatly on the alternative options for investment open to farmers. In Uttar Pradesh, many plots remain fallow and there remains significant unexploited groundwater irrigation potential. Farmers with saline lands often find it more profitable to sink a well on a neighboring plot, for example, or to increase fertilizer applications on another plot, than to invest in reclaiming a salt-affected plot. In Punjab and Haryana, on the other hand, groundwater potential is almost completely exploited, and there are relatively few other investment opportunities to increase land productivity. As a result, investment in reclamation of salt-affected soils is quite high. Joshi estimates that 42% of saltaffected plots have been reclaimed in Punjab and 36% in Haryana, but only 11% in Uttar Pradesh.

Joshi recommended that credit programs are needed to encourage private land reclamation. Drainage, meanwhile, requires government assistance because it must be addressed on a scale larger than an individual farm or even a village. Efforts are needed to encourage farmers to act collectively to maintain drainage arrangements.

Groundwater Degradation

Groundwater is another resource that, in many places, is inextricably linked to rainfed agriculture. This is especially so in semi-arid areas with unreliable monsoon rains. Although access to groundwater immediately converts a rainfed agricultural plot to an irrigated plot, it still has important implications for dryland farming because virtually all farmers with wells also operate rainfed plots, and they allocate their time and financial resources between irrigated and rainfed agriculture. Likewise, through joint ownership of wells and groundwater market, the percentage of farmers with access to some irrigation greatly exceeds the percentage of area irrigated (Mehra, 1995; Shah, 1993). Walker and Ryan (1990), in fact, cite farmers' interest in irrigation as a major constraint to adoption of improved dryland agricultural practices: farmers would rather allocate their money to irrigation wells than dryland technologies. Although wells have only moderately high expected rates of return (Pender, 1993), farmers prefer to invest in them in part because of the relatively stable production obtained through irrigation.

Groundwater is a difficult resource to manage, especially in hard-rock areas characterized by low groundwater recharge and small aquifers whose boundaries are not known. Irrigation wells have existed for centuries in India, but they have only become widespread in the last couple of decades, as a result of easy access to electricity and diesel fuel. In some areas, groundwater exploitation has reached such an advanced stage that water tables are falling in some places by more than a meter per year, so that farmers have to deepen their wells every few years to retain access to water (Vaidyanathan, 1994, cited by Repetto, 1994). In some areas, deep drilling has led to saltwater intrusion into aquifers, rendering water unfit for agricultural or domestic use.

Two factors have contributed to growing scarcity of groundwater in many areas in the last decade: low, flat rate prices for electricity to power wells, and the absence of property rights to groundwater. Regarding property rights, anyone in India who owns a plot of land has the right to sink a well and extract virtually as much water as they please. Aquifers, however, underlie a great many plots, so one person's pumping may reduce another's access to groundwater. Wealthier farmers have a competitive advantage over their less wealthy counterparts in this process, because wealthier farmers can more easily make the repeated investments to deepen a well. State governments have introduced some restrictions intended to prevent the digging of new wells in areas with too much pressure on groundwater, but as Repetto (1994) points out, such provisions only solidify the gains that wealthier farmers made by investing in wells first. Less wealthy, later adopters are unfairly excluded. This inequality is reinforced by subsidies for credit, fertilizers and other inputs that are used more intensively by farmers with irrigation (Repetto, 1994).

Shah (1993) suggests mechanisms to convert groundwater from an open access resource to a community-based common property resource with well-defined rights for local users, but numerous logistical difficulties pose a serious challenge to such efforts. Despite the difficulties, ultimately such an approach is probably the best hope for a property rights specification that generates an allocation pattern that is efficient, equitable and environmentally sustainable.

Most states have charged flat rates for power consumption since the early 1980. Flat rates are easier to administer than per unit pricing, but they have serious, negative implications for groundwater management in water-scarce areas. Under flat rate power tariffs, there is no relationship between the quantity pumped and the tariff paid, so well owners have an incentive to pump water until the average product is practically zero. Farmers have responded Panchayats 9 could play a role in groundwater management if they are strengthened. Panchayats are discussed further in section 7. predictably, planting water-intensive crops such as sugarcane and paddy. State electricity boards have resorted to quantitative rationing to limit pumping, and higher tariffs to industrial users that effectively subsidize agricultural use. The results are scheduled and unscheduled power cuts for industrial, agricultural and residential users; structural imbalances in the allocation of power between sectors; and continuing deficits for most state electricity boards (Kerr et al 1996).

Some analysts have suggested that higher flat rate tariffs will solve the problems facing electricity boards. Kerr et al (1996), however, demonstrate that even high flat rates do not create an incentive for farmers to reduce pumping, because once the fee is paid there is still no relationship between the amounts pumped and paid. The high fixed fee may also be inequitable, because it may constrain cash-constrained farmers from purchasing an electricity connection in the first place.

Property rights and power tariff problems may have to be solved jointly, through collective property rights to groundwater. As mentioned above, Shah (1993) suggests some cooperative approaches to groundwater management. Kerr et al (1996) suggest that property rights and power tariffs could be village-based, with laws to designate the relative rights and duties of farmers who own wells and those who don't. This would be a second-best solution that ignores the fact that aquifer boundaries and village boundaries do not match, but it may prove to be a pragmatic compromise given the difficulty of mapping aquifers and organizing people around aquifer boundaries.9 So far we have focused on the problem of groundwater scarcity in relatively dry areas. In many canal command areas, on the other hand, environmental problems associated with groundwater paradoxically result from its underexploitation. This is because in canal command areas, pumping groundwater has the beneficial effect of reducing the threat of waterlogging and salinization by lowering the water table, which tends to be high in surface irrigated areas. Farmers in such areas have less incentive to develop groundwater resources because they already enjoy virtually free access to surface water. In these areas, exploiting groundwater is associated with positive externalities and thus should be subsidized by the government. Low, flat rate prices make sense under these circumstances (Shah 1993). Finally, Repetto (1994) points out that groundwater remains underexploited in higher rainfall areas of eastern India. This is attributed to the lower marginal benefits of irrigation in such areas than in semi-arid regions; the poor state of rural electrification and other infrastructure that is conducive to agricultural intensification; and complex tenure relationships that inhibit expensive, long term land improvement investments.

Degradation of Uncultivated Lands

The poor condition of India's village forest and pasture lands evoke stark images of the "tragedy of the commons." Jodha (1992) documents the breakdown of traditional mechanisms to govern use of common lands and the severe consequences in terms of decline in their area, productivity and employment generation. Area under common lands declined by 30-50 percent in 8 semi-arid states between 1950 and 1980. Bentley (1984) estimates that over 80% of India's 123 million hectares of uncultivated lands produce 20% or less of their biological potential. He argues that low productivity is particularly acute on the common lands that make up most of this area.

While about 23% of India's territory is officially classified as forested, the area actually covered by trees is no more than about 10-12% (Bentley, 1984; CSE 1982). Some forest lands have been converted to grazing pasture, but these too are unproductive (Gadgil, 1982). The resulting shortage of biomass has significant implications for the development of rainfed areas. One well-known problem is that villagers have to look for other fuel sources in addition to wood to cook meals and heat water. Chambers et al (1989) cite UNDP figures showing that the real price of firewood in rural areas doubled between 1973 and 1975. As a result, cow dung has a high opportunity cost for use as fuel, so its application to crops has declined (Motavalli and Anders 1991). Farmers correctly perceive that manure has residual productivity effects lasting up to three years, but many farmers do not apply manure so frequently. In dryland areas, irrigated plots and the most productive dry plots receive manure more manure than less productive dry plots. Motavalli and Anders found that about 70% of irrigated plots received manure at least once every three years; the figure for dryland plots was less than 20%. Biomass shortages also limit farmers' ability to retain stubble on crop lands during the dry season. Farmers cut as much stubble as possible to use as feed or fuel, and grazing animals finish what is left. The bare soil that remains is highly prone to erosion by early rains. Cogle and Rao (1993) report that applying straw mulch to alfisols greatly raises yields by increasing infiltration and reducing erosion, and that this practice would pay for itself within a few years. It is not known if farmers will be willing to adopt this practice, however, given the high, immediate demand for fodder and the scarcity of capital with which to make long term investments.

The decline in forest cover has other negative implications for rural areas. Gupta (1982) estimated that for every person directly employed in forestry, four are employed indirectly though forward linkages. Gupta (1982) also estimated that nontimber forest products generated 2 million man-years of employment in India, and that the number could be more than doubled if markets were better developed. Bentley (1984) pointed out that a much larger increase could be realized if uncultivated lands were managed for higher productivity. Jodha's (1990) finding that common lands in 24 villages in 8 semi-arid states generated an average of about 150 employment days for poor families further demonstrates their employment potential. In these same villages, the poorest families derived as much as 25 percent of their incomes from common lands.

Numerous causes underlie the decline in production of common lands; we mention them only briefly here. Repetto (1994) (citing Jodha 1992 and Agarwal 1992) outline some of the traditional mechanisms for managing village common lands in pre-colonial times and their emphasis on sustainability. The British Colonial administration usurped the ownership of these lands, managing them as a source of government revenue and natural resources to support the colonial economy (Bentley 1984, Gadgil and Guha, 1992). An adversarial relationship between villagers and Forest Departments developed as early as the late 19th century, when villagers were deprived of their traditional rights to natural resources that were an integral part of their local economy (Gadgil and Guha, 1992). Independence did not change this situation as the new government retained the same approach to forest management. Government policies also reserved many areas for forest-based industries, which received access to forest products at heavily subsidized rates and with little or no obligation to manage forest resources on a sustainable basis. People living in forest areas had little stake in the long term health of forest resources because they would be excluded from the benefits. As a result, they would intrude on the forests when they could, illegally taking what resources they could.

On common lands not managed by Forest Departments, the colonial administration nurtured relationships with large landlords (zamindars) who took responsibility for enforcing restrictions on the use of common lands. The zamindari system worked from the perspective of managing natural resources, but it was feudalistic in nature, with highly inequitable distribution of benefits. After Independence, the new government stripped zamindars of their power and with it, the effective systems for restricting the exploitation of common lands. State governments attempted to take responsibility to manage common lands but with poor results (Bentley 1984).

Responses to the Declining Productivity of Common Lands

Declining productivity of common lands has stimulated two kinds of responses. One response is action by government and nongovernment organizations to develop special programs to manage forests and pastures, and the other is a spontaneous response by rural people to either manage these lands differently or develop new ways to produce the goods that the commons no longer supply. We discuss each of these in turn. Government initiatives. Special programs come in many forms.

Among the largest has been the government's effort to promote social forestry, in which the state Forest Departments involve rural people to manage trees, rather managing trees to the exclusion of rural people. Table 7.5 shows that between the 1950s and 1980s, the government both greatly increased its emphasis on tree production and changed its approach to doing so. Under social forestry, the aim was to grow trees on village common lands, roadsides, and private plots, rather than in the reserved forests where people were not allowed.

Table 7.5 Forestry expenditures in the 1st and 6th plans

	Social Forestry		Reserved Forests	
	ha (mill)	percent	ha (mill)	percent
1st Plan (1951-56)	15	29	37	71
6th Plan (1980-85)	1524	71	624	29

Source: Bentley, 1984

In the initial efforts to promote social forestry, production was concentrated on private lands, while common lands remained unproductive, with low survival rates (Bentley 1984). Social forestry projects often worked with private farmers simply because it was easier to grow trees on private than common land. There was nothing particularly "social" about such forestry initiatives. Following this experience, more recent projects working on common lands have tried to organize and motivate people to act collectively to protect trees. These projects have had mixed success, in part because of differences in the way they are managed (Hinchcliffe et al, 1995; Kerr and Pender, 1996a, but also because of differences in communities: some communities may be more willing to engage in collective action than others (Wade, 1988).

Legal rights of villagers may be critical to the success of efforts to develop common lands. There is growing appreciation of the fact that under Indian forest laws, local people have had very limited rights to the products of forest lands, and this has reduced their incentive to protect them. This realization led to the development of Joint Forest Management, an arrangement that represents a compromise between state ownership of forest lands and increased access rights for local people. Under Joint Forest Management, villagers receive a certain percentage of the proceeds of timber sales; they also own the rights to all nontimber forest products. The logic is simple: villagers will have more incentive to protect the forest if they own a share of the benefits. Institutions for collective action within villages still need to be strengthened, but Joint Forest Management is a step that will increase the returns to collective action. Joint Forest Management is a recent initiative; it will take more time to judge its performance.

The new Panchayat Raj law is a more broad-based initiative to transfer rights and responsibilities for various aspects of the rural economy, including natural resource management. Although this movement is in its infancy, in some respects it represents a return to precolonial systems of village level autonomy. Villagers will determine how and where to invest public funds for development of their economy, and they will make and enforce rules for managing many natural resources. This will provide an interesting test of the hypothesis that villagers manage common property natural resources poorly because they are alienated by laws that limit the benefits they can obtain from them. The Panchayat Raj is discussed further in section 9.

Rural people's actions. As mentioned above, rural people have also acted spontaneously to cope with the reduced area and productivity of common lands.

These spontaneous actions receive less attention than special government and NGO projects, but they may be more significant because they represent endogenous change with concrete results. Farmers have shifted their cropping patterns in response to the decline of government and common lands. Trees, for example, are increasingly cultivated by private farmers for sale to industrial users in large cities. Table 7.6 shows how timber prices rose between 1970 and 1987 relative to other prices; it is easy to see why farmers responded to these prices by growing timber. Eucalyptus plantations surrounding Bangalore, for example, bear testimony to this change in the sources of timber.

Private tree cultivation has generated so much supply in north India that markets crashed, leaving tree growers with large losses (Saxena 1990). Farmers' choice of sorghum varieties is a case in which farmers have changed their cropping patterns at least partly in response to declining fodder production on common lands. Kelley et al (1993) document farmers' shift toward sorghum varieties that produce more straw, relating it to an increase in the price of fodder relative to the price of food grain. In this way, crop lands substitute for common lands to produce fodder.

Shepherds in Mahbubnagar District of Andhra Pradesh have taken a different approach to changing production patterns on crop lands. Common grazing lands that once supported herds of sheep now have declined in area due to land distribution programs, and they contain little besides boulders and a few shrubs. Shepherds have responded by forming informal cooperative groups of about 5 farmers and 500 sheep and leasing or buying discontinuous plots of marginal crop land owned by high caste farmers with excess land. The shepherds manage these lands as grazing pastures. Cooperation enables these shepherds to exploit scale economies in managing larger herds and following rotational grazing on the large pasture area. A critical factor that enables this system to work is the high price of mutton, which makes it profitable to lease in land.

Table 7.6 Rising prices of timber and overall agricultural prices

	General Index of Wholesale Prices	Index of Wholesale Agricultural Prices	Index of Timber Prices
1970-71	100	100	100
1975-76	173	157	178
1980-81	257	211	407
1981-82	281	237	556
1982-83	288	248	740
1983-84	316	283	811
1984-85	338	303	946
1985-86	358	310	821
1986-87	377	330	866
July, 1987	401	368	945

Source: Chambers et al, 1989

But the initiative of the shepherds is what has enabled this resource management system to evolve (personal observation and personal communication with Berend de Groot, Director, Indo-Swiss Dairy Development Project, Hyderabad).

In addition to shifting cropping patterns, villagers in many places have changed the way they manage common lands. Privatization of the commons has occurred in many places. In most areas the privatization process reflects the problem more than a response to problems of common land management, as populist government programs distribute ever more marginal lands to landless constituents (Pender and Kerr, 1996). In some areas, however, villagers have taken the initiative to privatize common pasture lands informally. Kerr and Pender (1996a) document this process for the case of grazing pastures in southeastern Rajasthan; they find that both visual evidence and most biological indicators suggest private pastures are better managed than their common counterparts.

In other cases, the outcome of greener pastures or more productive forests may be attained through alternate approaches. Anecdotal information abounds of cases of successful collective action to manage village common lands productively. Honey Bee (1995), for example, cites the case of Mr. Balvantsinh in the village of Takhua, Banaskantha district, Gujarat who, alarmed by the degradation of common lands in his village, single-handedly mobilized his fellow villagers to protect common lands. In this particular case, linking the management of the commons to religious practices and associated duties was a critical element in making the management system work. Though cases such as this one appear to more of an exception than a rule under current conditions, they do provide evidence that collective action can work. More significantly, they demonstrate the capacity of local institutional innovation to develop a system that is appropriate to the prevailing circumstances. Legal reforms that guarantee villagers greater benefits from protecting the commons will support such efforts. The Panchayat Raj and Joint Forest Management are prime examples.

These examples of spontaneous responses to the problems of declining productivity of common land resources demonstrate the need for caution when documenting resource degradation and estimating its impact on the rural economy. These few examples, some isolated and others widespread, probably have innumerable counterparts that can be observed easily if the effort is made to search for them. Declining productivity of the commons does not occur in isolation; rather, rural people devise mechanisms to cope with the consequences 10The coefficient of variation is equal to the standard deviation divided by the mean. of reduced productivity of common lands. Simple indicators of the consequences of the declining commons may be misleading as a result. This is not to suggest that some people are not hurt by the declining productivity of the commons; Jodha (1990), for example, shows that the poorest people suffered the most from the declining commons. But there is a need to take the next step of finding out how such people have responded to changing circumstances. If the rural economy has changed significantly, they may have replaced their dependence on the commons with some other livelihood strategy. The low-caste shepherds of Mahbubnagar, mentioned above, initially suffered from the decline of common grazing pastures, but now they are among the wealthier groups of villagers (personal communication, Y. Mohan Rao, Senior Research Associate, ICRISAT).

8

RISK AND RAINFED AGRICULTURE

Rainfed agriculture in semi-arid or sub-humid regions is generally risky and unstable. Rainfall in the semi-arid tropics is not only low, but also unreliable, with a higher CV than in more humid areas. Weather-related risk places hardship on people in these areas, and it may constrain adoption of more productive agricultural technology. In this section we review the evidence on instability and its consequences, and we discuss some mechanisms to alleviate the problems.

INSTABILITY OF RAINFED VS. IRRIGATED AGRICULTURE

Agricultural yields are more unstable in rainfed areas than where irrigation is assured. But if irrigation water supplies fail, then irrigated agricultural area can fall sharply, leading to even more unstable output than under rainfed agriculture. Dhawan (1988a) used the coefficient of variation10 (CV) as the indicator of instability of production, yield and area cultivated for irrigated and rainfed food grains. Using national data, Dhawan found that the CV of output of unirrigated food grains for the period 1970-83 was 11.4%, significantly higher then the 6.4% for irrigated food grains. Decomposing these into output and yield instability, the CVs of yield were 9.2% and 5.3% for rainfed and irrigated food grains, respectively, and for area they were 3.1% and 2.3%, respectively. This suggests that yield and area moved together to increase output instability. Shah and Sah (1993) also found higher yield variations in rainfed than irrigated crops. They calculated CVs ranging from 36% to 70% for yields of rainfed food grain crops and 9%- 26% for irrigated food grains. They did not examine how these yield variations translated into production instability. The larger variations found by Shah and Sah most likely result from their small sample size of a few rainfed and a few irrigated districts over 10 years in the state of Gujarat. Dhawan's study was based on nationwide data, which is likely to be more smooth. Walker (1989b) found similarly large fluctuations in the yields of both rainfed and irrigated crops in a village level study in Maharashtra and Andhra Pradesh. This study was based on 40 households in each of three villages over 10 years, so it is not expected to be smooth. For irrigated paddy in one Andhra Pradesh village, the mean CV was 31% between 1975-76 and 1983-84, and for several rainfed crops in three villages in Andhra Pradesh and Maharashtra, the mean CVs ranged from 44% to 69%. In tank irrigated areas of south India, Hazell and Ramasamy (1991) found that paddy production fell 50% in their study villages in Tamil Nadu in the 1982-83 drought; virtually all of the decline resulted from reduced area planted. Bidinger et al (1990) found an even greater decline in paddy area in their study village in Andhra Pradesh in the 1985-87 drought, with little change in irrigated yields.

INSTABILITY ASSOCIATED WITH IMPROVED AGRICULTURAL TECHNOLOGY

There has been much concern over the possibility that improved crop varieties are associated with increased output instability in addition to higher yields. If this is the case, it could deter adoption of more productive technology and hence retard agricultural development.

Instability at the Farm Level

In discussions of farmers' adoption of HYVs, it is often assumed that new seeds have more volatile yields. Indigenous landraces often have the characteristic of being fairly unresponsive to improved management and increased inputs, but also robust in the face of unfavorable agroclimatic conditions, such as drought. HYVs, on the other hand, often are characterized as being highly responsive to management and inputs, but very low yielding in the event of bad weather. If this is the case, then plant breeders must breed on the basis of one strategy (high average yield) for high potential areas and another (risk minimization) for low potential areas.

Research on pearl millet, which is commonly grown in unfavorable agroclimatic areas, suggests that the above characterization of traditional varieties and HYVs is not necessarily correct. Witcombe (1989), in a study of Pakistan and India, found that good performance of a particular millet seed over all environments appears to indicate good performance in environments of low potential. In three years out of his four year study, the highest yielding entry across all environments was also one of the two highest yielding entries in the lowest yielding environment. These results suggest that the typical plant breeder's strategy of selecting among the highest yielding seeds across all environments is satisfactory. Farmers who adopt HYVs do not necessarily subject themselves to greater risk of catastrophic loss in the event of bad weather.

Farmers' Strategies to Reduce and Cope with Risk. Walker and Jodha (1986) point out that dryland farmers have various methods to reduce their exposure to crop production risk. Cultural practices play an important risk-reducing role; they include planting different crops with relatively low covariate yield (either in an intercrop or on separate fields); diversifying spatially by operating multiple plots with different environmental characteristics; and staggering planting dates in the face of variable rainfall patterns. Sharecropping is a common tenancy arrangement that distributes risk between the tenant and landlord. Many farmers have multiple sources of income, reducing risk if they have low covariation.

Farmers also have various mechanisms to cope with risk that they cannot eliminate. For example, they can borrow from local stores or money lenders, draw down food stocks or savings, sell assets, obtain transfers from relatives, participate in government relief programs, or migrate. Most of these options are not particularly desirable; for especially poor people they can be quite devastating: selling assets or going into debt may make a family permanently worse off even after drought is over. Jodha (198_) indicates that some families will reduce their food consumption as much as possible before parting with their assets; this has obvious negative short term health implications that are particularly severe for those who consume only minimum requirements to begin with.

Effect of Risk on Technology Adoption. While weather-related risk undoubtedly presents great hardship for a very large number of people, Walker (1989b) indicated that it may not be as important as generally believed in adoption of new technologies in rainfed areas.

The tradeoff between expected income and the variance of that income suggests that given farmers' measured risk preferences, the overall effect on adoption decisions is modest. Also, Walker stressed that yield instability does not translate into major variability in income due to farmers' mechanisms to absorb risk. As mentioned above, multiple sources of income and diverse cropping patterns mean that yield variability of one crop only affects a portion of the income from crops. Equally important, in many cases farmers can adjust the area under each crop depending on the weather at the start of the season. For example, dryland rabi (postrainy season) sorghum farmers in black soil areas know at the start of the season how much moisture is available and adjust their cropped area accordingly. Similarly, castor farmers in Andhra Pradesh know that pest attacks are more prevalent when the rainy season begins late, so they plant less castor. And farmers with irrigation from tanks or wells know roughly how much water will be available in the postrainy season, so they adjust planted area accordingly. Hazell and Ramasamy (1991), for example, found that sharply reduced paddy production in North Arcot, Tamil Nadu, in the drought year 1982-83 resulted from a fall in area, not yield. Not surprisingly, Walker (1989b) also found that variations in cultivated area This is why SWC programs are sometimes referred to as "drought-proofing" measures.

exceeded those in yield. He also found that given farmers' diversified farming and livelihood strategies, the large variations in yield translated into only small variations in income. The variations were so small that even if plant breeders could develop varieties with perfect stability (zero variation), the contribution to household income stability would amount to less than 1% for most of the crops studied, and a maximum of 2.9% for paddy (which, paradoxically, had the least yield variability to begin with).

Walker's (1989b) results demonstrate the importance of looking beyond yield variability of a single crop to variations in all the crops in a given household, village or region, and beyond crop variation to income variation. From this perspective, yield instability does not appear to be a major determinant of adoption, and hence not a top priority for plant breeders. He pointed out (in his 1989a study) that policies related to international trade and storage between surplus and deficit years can be more cost-effective in coping with increasing yield instability.

Soil and water conservation (SWC) investments also are associated with a variety of risks. First, erosion itself is a matter of risk. For example, some plots may be at risk of productivity loss from continuous, gradual erosion, but others may be more susceptible to significant erosion only in the event of a once-in-five-years or once-in-fifty-years storm. For such lands, in normal years there is no gain from investments to reduce erosion, but in exceptional years the gain — actually the avoided loss — may be quite high. Second, often SWC practices serve to conserve soil moisture as well as reduce erosion. Soil moisture retention is more likely to offer immediate, productivity-increasing benefits than erosion prevention. However, these benefits may exceed the costs only when rainfall is unusually low or unevenly distributed, so that moisture stress constrains productivity.12 In a good rainfall year, on the other hand, short term gross returns to increased moisture retention may be low or zero, and in a very high rainfall year they may even be negative if they lead to waterlogging. Little information is available regarding the impact of risk on Indian farmers' SWC investments. As mentioned above in section 5, most SWC or watershed projects are so heavily subsidized that risk is not a factor. The little research conducted on adoption outside of such projects did not address the issue of risk.

Instability at the Aggregate Level

Ramakrishna (1993) used a log function to calculate CVs to compare instability of production, yield and area cultivated for cereals between the pre-green revolution and green revolution periods. The CVs are shown in table 8.1.

Ramakrishna found that for food grains, output was more stable during the green revolution than before. Yields were about equally stable between the two periods, and area became more stable. For other crops, on the other hand, Ramakrishna found that output and yield instability both increased during the green revolution, while area again became more stable. Combining food grains and other crops, output and yield instability increased slightly during the green revolution, while area instability was constant. On the whole, Ramakrishna's data suggest that the green revolution had little impact on agricultural instability.

Hanumantha Rao (1994) took a slightly different approach and found different results. Instead of taking CVs of output levels, he took the standard deviation of annual output growth rates as a measure of instability. He found that the standard deviation of output growth of food grains was 8.1 in 1950-51 but 11.4 between 1968-85, and took this to indicate that output instability increased with the green revolution. He also found that the standard deviation rose from 9.4 in the first decade of the green revolution to 11.8 in the second decade. Hanumantha Rao did not provide detailed figures regarding yield and area instability. His figures do not really offer a good comparison to those of Ramakrishna, because his pre-green revolution figures are based only on a single year. Also, it is difficult to compare the magnitude of Hanumantha Rao's figures (expressed as the standard deviation of output growth rates) to Ramakrishna's (expressed as the CV of the level of output). Since we do not know the growth rates from which Hanumantha Rao calculated the standard deviation, we cannot relate his standard deviations to Ramakrishna's CVs. Hazell (1982) examined cereal crop production between 1954-55 and 1964-65 (before the green revolution) and 1967-68 and 1977-78 (during the green revolution, after the introduction of high yielding varieties).13 He found that the

CV of production increased by about 50%, from 0.04 to 0.059 between the two periods. Hazell hypothesized that if the increased instability were due to HYVs, variances in production within states would have to rise. However, he found that changes in yield covariances were much more important than changes in yield variances. Only about 18 percent of the increase in variance of total cereal production resulted from changes in crop production variances; the remaining 82 percent was explained by changes in covariances; interstate covariances within crops contributed 41 percent to the change in variance in total cereal production. As a result, Hazell concluded that HYVs were probably not the primary cause of increased variability. In a later paper, Hazell (1984) suggested that HYVs could possibly affect yield covariances of maize in India and the United States, because the narrower genetic base of improved varieties would make them susceptible to common yield inhibitors such as pests and diseases. This would then be a contributing factor to greater variability in national ceral production.

Walker (1989a) studied changes in yield and output variability of sorghum and pearl millet in India resulting from the spread of HYVs. This study followed Hazell's 1984 study relating the rising covariance of yields between states to the spread of HYVs; it used districtlevel data for the 48 largest sorghum districts and the 40 largest pearl millet districts. Walker found that variability had indeed increased significantly between the period 1956-57 to 1967- 68 and 1968-69 to 1979-80, from 8% to 16% for sorghum and 11% to 34% for pearl millet. He also found that covariance of yields across districts was by far the most important factor in overall output variance of sorghum and pearl millet. Walker examined several possible causes of increased covariance, including 1) changes in rainfall covariance, 2) changes in irrigated area, and 3) adoption of HYVs. He found evidence that provided weak support for each of these possible sources of covariance. However, he also stressed that the contribution of HYVs to increased instability is dwarfed by their contribution to productivity, so he did not recommend changes in existing breeding strategies that develop HYVs for adoption over a large area.

Walker's (1989a) study helps clarify that numerous factors can contribute to changes in stability. Researchers who find that stability increased or decreased with the green revolution should hesitate before proclaiming that HYVs were the cause. Changes in cropping patterns, fluctuations in weather, and the quality of land on which a particular crop tends to be planted all can have an impact on the variations in yield. More importantly, variations in area may either counteract or reinforce those in yield; it is important to know the composition of output fluctuations between yield and area variations. The studies by Hazell and Walker also highlight the need to distinguish between variability at the regional or national levels from farm or field level risks. Because of the dominance of covariance relations in aggregate production data, variability can increase at the aggregate level even while farm level variability changes little or not at all. Although diets were stable, health problems increased due to a shortage of clean drinking water in the village resulting from the drought.

DROUGHT RISK

The preceding discussion on output instability has underemphasized the risks associated with the specific problem of severe, prolonged drought. When severe drought occurs over a wide area, yield and income risks can be particularly great. In Hazell and Ramasamy's (1991) study of North Arcot district, Tamil Nadu, for example, they found that the 50% decline in paddy production in their study villages was replicated throughout the district and beyond.

Incomes fell by 50% on average as even nonfarm incomes fell sharply. Consumption expenditure fell by 50% on average, and diet quality deteriorated. Clearly, the effects of drought were covariate across villages and sectors. This problem of covariate risk is particularly challenging because poor performance of one income source, or one location, cannot necessarily be compensated by better performance in another. As a result, people may face severe hardships in the event of drought.

In the Bidinger et al (1990) study, on the other hand, despite a similar 50% decline in income, dietary intake did not change compared to the pre-drought situation.14 People maintained their dietary intake through increases in temporary migration and consumption credit, and government rice subsidies that kept prices low and stable, enabling people to translate meager incomes into normal diets. The number of people who migrated only increased by about 10%, but they stayed away much longer on average. Interest rates did not show the expected increase as demand for consumption credit rose, even though the bulk of borrowing was done through the informal village credit market rather than the formal banking system, which contributed little. Distress sales of land and livestock were rare, but two families did sell all their assets and move permanently to become urban laborers.

GOVERNMENT INTERVENTIONS TO MANAGE RISK

Severe droughts impact negatively on most rural households simultaneously and are therefore difficult to manage through traditional risk sharing and coping strategies. As a result, government policy may sometimes play an important role in helping farmers manage risk. While some government interventions are well-established, other ideas remain relatively untested. Likewise, even where farmers and rural communities have developed effective mechanisms to manage production risk, it is not clear what the costs of these mechanisms are in terms of reduced production efficiency. Perhaps market-based or government-sponsored alternatives can be introduced that protect farmers from the effects of risk but without requiring diverse income sources and fragmented agricultural holdings. Anderson and Hazell (1994) point out that more information is needed about the costs of these risk-reducing and coping mechanisms. In this section we briefly review government measures to manage risk and discuss some additional possible government approaches.

Most of the existing drought management efforts were discussed above, in section 5. These are poverty allevation programs to cope with the effects of drought, namely employment programs and food subsidies. There is no need to repeat the discussion of these programs here. Bidinger et al (1990), stress the importance of food subsidies in mitigating the effects of drought in their study, but they also point out that a timely public works program could have prevented much of the unemployment and debt experienced by laborers in the village.

Crop insurance is provided by the public sector in many countries. The impetus for such programs often originates in governmental concern about catastrophic risks such as drought, or the desire to reduce the incidence of loan defaults to banks. With few exceptions, the financial performance of public crop insurers has been ruinous (Hazell 1992). To be financially viable without government subsideis, an insurer needs to keep the average value of its annual outgoings—indemnities plus administration costs—below the average value of the premiums it collects from farmers. In practice, many of the larger insurance programs pay out $2.50 or more for every dollar of premium they collect from farmers. The difference is paid by governments, at costs varying from $10 to $400 per insured hectare. Even at these levels of subsidy, many farmers are still reluctant to purchase insurance. As such, many crop insurance programs are compulsory, either for all farmers growing specified crops or for those who borrow from agricultural banks. The primary reason for the high cost of public crop insurance schemes is that they invariably attempt to insure risks that are prone to severe moral hazard problems, whereby farmers have incentive to lie about their production levels, or to allow their crops to fail in order to receive insurance payments (Hazell 1995a). These risks include many climate, disease and pest risks that are difficult to quantify and assess, and whose damage can be influenced by armers' management practices. The problem is aggravated by a common practice of insuring "target" yields rather than compensating for actual losses. But this is not the only reason for failure.

Another overwhelming factor is the incentive problem that arises oce the government establishes a pattern of guaranteeing the financial viability of an insurer. If the insurance staff know that any losses will automatically be covered by governmetn, they have little incentive to pursue sound insurance practices when setting premiums and assessing losses. In fact, they may find it profitable to collude with farmers in filing exaggerated or falsified claims. Yet another common reason for failure has been that governments undermine public insurers for political reasons. In Mexico, the total indemnities paid has borne a strong statistical relationship with the electoral cycle, increasing sharply immediately before and during election years, and falling off thereafter. In the USA, the government has repeately undermined the national crop insurer (FCIC) by providing direct assistance to producers in disaster areas. Why should farmers purchase crop insurance against major calamities (including drought) if they know that farm lobbies can usually apply the necessary political pressure to obtain direct assistance for them in times of need at no financial cost? Another reason for their high cost is that crop insurers tend to be too specialized, focusing on specific crops, regions and types of farmers, particularly when the insurance is tied to credit programs designed to serve particular target groups identified by the government. Without a well-diversified insurance portfolio, crop insurers are susceptible to covariability problems, a face the prospe t of sizable losss in some years. Since public insurers are rarely able to obtain commercial reinsurance or contingent loan arrangements, thsi specialization increasestheir dependence on the government.

Crop insurances is provided by the public sector in many countries. The impetus for such programs often originates in governmental concern about catastrophic risks such as drought, or the desire to reduce the incidence of loan defaults to banks. With few exceptions, the financial performance of public crop insurers has been ruinous (Hazell 1992). To be financially viable without government subsidies, an insurer needs to keep the average value of its annual outgoings — indemnities plus administration costs — below the average value of the premiums it collects from farmers. In practice, many of the larger insurance programs pay out $2.50 or more for every dollar of premium they collect from farmers. The difference is paid by governments, at costs varying from $10 to $400 per insured hectare. Even at these levels of subsidy, many farmers are still reluctant to purchase insurance. As such, many crop-insurance programs are compulsory, either for all farmers growing specified crops, or for those who borrow from agricultural banks. The primary reason for the high cost of public crop-insurance schemes is that they invariably attempt to insure risks that are prone to severe moral hazard problems whereby farmers have incentive to lie about their production levels, or to allow their crops to fail in order to receive insurance payments (Hazell 1995a). These risks include many climate, disease and pest risks that are difficult to quantify and assess, and whose damage can be influenced by farmers' management practices. The problem is aggravated by a common practice of insuring "target" yields rather than compensating for actual losses. But this is not the only reason for failure.

Another overwhelming factor is the incentive problem that arises once the government establishes a pattern of guaranteeing the financial viability of an insurer. If the insurance staff know that any losses will automatically be covered by government, they have little incentive to pursue sound insurance practices when setting premiums and assessing losses. In fact, they may find it profitable to collude with farmers in filing exaggerated or falsified claims.

Yet another common reason for failure has been that governments undermine public insurers for political reasons. In Mexico, the total indemnities paid has borne a strong statistical relationship with the electoral cycle, increasing sharply immediately before and during election years, and falling off thereafter. In the USA, the government has repeatedly undermined the national crop insurer (FCIC) by providing direct assistance to producers in disaster areas. Why should farmers purchase crop insurance against major calamities (including drought) if they know that farm lobbies can usually apply the necessary political pressure to obtain direct assistance for them in times of need at no financial cost? Another reason for their high cost is that crop insurers tend to be too specialized, focusing on specific crops, regions and types of farmers, particularly when the insurance is tied to credit programs designed to serve particular target groups identified by the government. Without a well-diversified insurance portfolio, crop insurers are susceptible to covariability problems, and face the prospect of sizable losses in some years. Since public insurers are rarely able to obtain commercial reinsurance or contingent loan arrangements, this specialization increases their dependence on the government. Public crop insurers also tend to have high administration costs. This is partly because they often insure small-scale farmers, but also because crop-insurance work is very seasonal, and the absence of a well-diversified portfolio means that staff and field equipment are underemployed for significant parts of the year.

There is no convincing evidence that public subsidization of crop insurance has been socially beneficial. Indeed, social benefit-cost analyses of the Mexican and Japanese schemes show negligible social returns in relation to their high costs (Bassoco et al. 1986, Tsujii 1986). Nor is there much evidence that it has increased agricultural lending or benefited agricultural banks. In a rare study, Pomareda (1984) compared the performance of insured and uninsured loans in the portfolio of the Agricultural Development Bank (BDA) of Panama. Insured loans had slightly higher and more stable returns than uninsured loans. They were also repaid and cleared from the books closer to their expected duration. But the overall gains to the Bank were modest, and could have been achieved more easily at no cost to the government simply by allowing a 2 percent increase in the interest rate that BDA charged its borrowers. This would also have been cheaper for the borrowers than the premium rates they paid for the compulsory insurance.

In order to overcome the major problems associated with crop insurance, and to substantially reduce its administration costs, several authors have proposed area-based yield insurance (e.g. Halcrow, 1948; Dandekar, 1977; Miranda, 1991). Under this proposal, the crop yield for a *homogeneous region* is insured, and all insured farmers in the region pay the same premium and receive the same indemnity. Indemnities are paid whenever the average yield for the region falls below some critical level irrespective of the actual yields obtained by individual farmers. Premiums are calculated on the basis of year-to year variations in the average yield for the region, and would vary from one homogeneous region to another in accordance with differences in risk levels. This approach reduces moral hazard problems, and hence broadens the range of yield risks that can be viably insured. Moreover, since the premiums and indemnities are identical for all insured farmers in a region, it avoids the *adverse selection* problem. The latter refers to the situation in which farmers facing below-average risk tend to drop out of insurance programmes if they are charged premium rates based on average risk levels but are paid indemnities based on their own losses. Also, by eliminating the need for field inspections and loss assessments, the cost of administering an area-based scheme could be kept very low. Providing farmers pay their premium, it is not really necessary that they even grow the crop that they have insured.

Despite its appeal and its potential scope for reaching small-scale farmers, there are problems with the proposal. First, the insurance will be attractive to individual farmers only if their yields are highly correlated with the average yield for the region. Dandekar (1977) argues the homogenous areas can be defined in India in which inter-farm yield correlations are positive, but their is a growing body of micro-evidence showing that yield correlations between plots within the same village, or even within the same farm, are surprisingly low (e.g. Walker and Jodha, 1986). Small differences in ground contours, slope, and wind and sun exposure can lead to substantial differences in the yield damage caused by unfavorable climatic, pets and disease events, as can a few days difference in planting dates or the crop varieties grown.

Second, the scheme is subject to severe covariability problems. When the average yield is below the critical value in a region, all the insured farmers have to be compensated simultaneously. An unsubsidized insurer could hope to survive only if the scheme were to span a large number of regions with negative or positive but weakly correlated yields. The alternative of seeking commercial reinsurance seems unlikely given private insurers' reluctance to insure yields against a wide array of perils.

India introduced a national area-yield crop insurance scheme in 1985 (the Comprehensive Crop Insurance Scheme, CCIS), following a pilot phase from 1979 to 1984. The insurance is only available to farmers who borrow credit from financial institutions; and the indemnities are paid directly to the lending institution. Mishra (1994) provides evidence from Gujarat that the insurance did increase lending to small farmers, though repayment rates varied slightly. Unfortunately, the premium rates charged farmers are only a fraction of the rates needed to cover the indemnities paid, and even the small premiums charged (1-2% of coverage) are heavily subsidized by government. As a result, the financial performance of CCIS to date has been disastrous. The CCIS collected Rs 11,961 lakhs ($40 million) of premium during 1985-92, but paid Rs 90,163 lakhs ($303 million) indemnities over the same period (Joshi, nd). Considering that government subsidized about half the premium collected, CCIS paid Rs 15 for each rupee of premium collected from farmers. The insurance program will clearly need to be reformed if it is to become a cost-effective risk management policy.

Another variant of area-based insurance is regional rainfall insurance. In this case the insurance pays out whenever the average rainfall for a region falls below some critical value. Rainfall is easier to measure than average yield, and with modern satellite imagery it is now possible to assess the soil moisture content for a region with the minimum of field inspection. This information could be used to confirm rainfall readings, or even used directly as the insured peril.

Unlike regional yield insurance, rainfall insurance is not tied to the performance of specific crops, Since most farm families rarely depend on a single crop for their total income, the attractiveness of rainfall insurance should depend more on how the total insurance is limited to the occurrence of severe drought or floods, it is likely that in semi-arid or floodprone areas the indemnities would coincide with catastrophic income outcomes for many rural families.

Drought insurance could be marketed rather like lottery tickets, employing lowincome people to sell tickets on a commission basis. Unlike standard lotteries, however, all ticket holders would win a prize in a disaster year, bot no prize would be given in non-disaster years. There is no need to restrict the insurance to farm families, and all types of rural households might find it attractive. This is because a decline in farm incomes in drought or flood years usually leads to a sharp contraction in the rural non-farm economy, and the incomes of many workers and businessmen decline in tandem. Drought insurance faces the same covariability problem as regional yield insurance, but because it is limited to a specific weather peril, it might be much easier to obtain international reinsurance. This prospect could be enhanced if the scheme were run by a commercial bank or insurer within the country.

Schemes of this kind have yet to be tried, and without pilot projects it will be difficult to assess their potential value. Their cost-effectiveness would also need to be compared with alternative means of assisting vulnerable households, such as relief employment schemes, and targeted food rations or income transfers.

9

INFRASTRUCTURE, INSTITUTIONS AND POLICIES

This section briefly addresses the role of infrastructure, institutions and policies in agricultural development. Part of the reason they are presented together is that the three issues have some degree of overlap with each other. We discuss two kinds of infrastructure: physical and social. Sometimes the distinction between the two is not very clear. Physical infrastructure includes roads, irrigation facilities, electrification, banks, markets and other things. Banks and markets, however, also can be considered as social infrastructure; in fact markets are a social institution. Formal and informal cooperative societies may also be considered as social infrastructure, and so can educational facilities. Policies, meanwhile, cover a wide range of issues, including institutions.

INFRASTRUCTURE

Table 9.1 gives some indicators of the levels of infrastructure in rainfed and irrigated areas. Studies of various forms of infrastructure have confirmed its important role in promoting agricultural development. Subbarao (1985), for example, focused on infrastructure related to delivery systems for crop inputs such as fertilizers, seeds and pesticides. This study described input markets and, using state-level data, examined the relationship between infrastructural development and the availability of private input delivery systems. Subbarao found that rainfed agriculture is characterized by poorly developed roads and scarcity of educational, marketing and financial infrastructure. Not surprisingly, he found that the presence of such infrastructural facilities had a strong positive impact on provision of inputs by the private sector.

Table 9.1 Level of infrastructure in SAT districts of India

Variables	Unirrigated SAT	Irrigated SAT
Barah and Binswanger (1981):		
No. of regulated markets/100000 km^2	50.49	61.10
KM of roads/10 km^2 of geog. area	1.34	2.94
Area under HYV as % of gross cropped area	0.47	6.45
Subbarao (1985):		
NPK/ha of gross cropped area	2.74	6.50
Per capita financial infrastructure. (Rs.)	259.80	642.50

Binswanger et al (1993) studied the interactions among infrastructure and agricultural output and investment by government and farmers. Using district-level data from 85 districts in 13 states, they found that education infrastructure and rural banks played a strong role in farmers' investment and input and output decisions. State governments, meanwhile, invested in infrastructure on the basis of a district's agroclimatic potential, while banks located branches where both agroclimate and infrastructure were favorable. As a result, agricultural output was the result of a complex series of interdependent relationships. One important implication of these findings is that irrigated areas are likely to enjoy superior infrastructural facilities, augmenting their agroclimatic advantages over rainfed areas. Better-endowed rainfed areas, in turn, are likely to have better infrastructure than drier, less favorable rainfed areas. Several studies have estimated the elasticity of agricultural output with respect to various forms of rural infrastructure. These studies confirm the importance of infrastructure and show its complementarity to favorable prices. Table 9.2 shows some previous estimates. The table suggests that prices, roads, markets and primary schools have all been found to have relatively strong impacts on agricultural growth, although the range of finding is quite broad. These findings are not necessarily in agreement with those from the production function analysis in section 4 above. Also, there is no information about interactions among different types of infrastructure or between infrastructure and other factors. The analysis in section 4 also suffers from this limitation. More detailed analysis is needed in order to derive robust conclusions regarding priority areas for further investments to promote agricultural growth.

DECENTRALIZATION AND LOCAL GOVERNMENT

Decentralization has been proposed as a way to improve the quality of government investment in rural development infrastructure and improve rural governance in general. State governments undertake construction of physical infrastructure on a large scale, and they assist in promoting social institutions such as cooperative societies. Although local government exists in the form of districts, mandals (subdistricts) and panchayats (collections of a small number of villages), in most states local government bodies have no power. Investment and resource allocation decisions are made by the state government despite its remoteness from the rural people who will be affected; in recent years there has been increasing concern that state governments have performed poorly in this area. In response, new legislation known as the Panchayat Raj has been enacted to transfer much decisionmaking power from state governments to panchayats. Under this system, the panchayats will decide how to allocate development resources earmarked to them by the state. In most of the country it is too early to know how the Panchayat Raj will affect agricultural and rural development because it has not yet been instituted. In most states, ongoing legislative negotiations have delayed the start of the new approach to rural governance.

At least one state, West Bengal, has made progress in implementing the Panchayat Raj approach. In fact, decentralization in West Bengal precedes the formal declaration of the Panchayat Raj by a long time. Dasgupta (1995) points out that panchayat elections have been held regularly since 1975, and that half the state's development and poverty alleviation budgets are spent through the panchayats. They are dominated by poorer members of rural society. This contrasts with other states in which panchayats are occasionally suspended by the state government, control few resources, and/or are dominated by wealthier members of the area.

Dasgupta provides several qualitative indicators of the importance of panchayats in rural development. He suggests that development funds are better spent in West Bengal than in other states because panchayat members, who live in the villages and have personal relationships with their constituencies, have a better understanding of peoples' needs and a greater sense of accountability. In section 6 on poverty, we discussed the problem that employment programs may face difficulties in creating permanent assets because beneficiaries focus more on short term employment benefits than long term asset creation. If employment funds are channeled through panchayats with more accountability and a greater stake in creating productive assets for the community, there is a greater chance of establishing a real link between employment and asset creation.15 they would recognize this problem and either: 1) ensure that trees were protected, or 2) allocate employment resources elsewhere. Similarly, panchayats could play a critical role in defining and enforcing collective property rights to groundwater, as discussed in section 7.

As an example of panchayats' legitimacy among rural people in West Bengal, Dasgupta notes that they play a strong role in dispute resolution even though their actual legal power is weak. Historically village level disputes, such as over property boundaries, damages resulting from intrusive grazing, etc., were handled at the village level (Jodha 1980). In recent years, however, village level institutions have disintegrated, so in many states villagers settle disputes in distant, state-run courts of law that can take months or even years to reach decisions. This contrasts sharply with the major role of panchayats in settling disputes in West Bengal.

Dasgupta points out various weaknesses with the West Bengali panchayat system, but on the whole, it provides some indication that the Panchayat Raj system may make a strong contribution to political and economic development in the future. On the other hand, there is still a shortage of quantitative measurement of the positive effects of decentralization. Rao and Kalirajan (1995) propose to undertake a study in which they will use econometric modeling to test the relationship between decentralized governance, on the one hand, and various development indicators on the other. In addition to the consistently strong status of panchayats in West Bengal, Andhra Pradesh and Karnataka have also decentralized, though various changes in state governments have interrupted the panchayats from time to time. Rao and Kalirajan propose to use district level data for their analysis and use dummy variables to indicate the districts in West Bengal, Andhra Pradesh and Karnataka that have had functioning panchayats. The results of this study are not yet available, but the proposed approach suggests a useful way to address the issue. Nongovernment Institutions and Weak Local Government. The lack of strong, credible local government in most states has stimulated the growth of informal committees and cooperative groups that provide members with important economic and social services. As discussed in section 6 on poverty and section 7 on watershed management programs, recent years have seen the spread of informal village level committees and cooperative groups that provide their members with important economic and social services (Fernandez 1991, Parthasarathy 1994). Various NGOs are helping villagers organize themselves into self-help groups that focus on issues ranging from credit to health to watershed management. Informal thrift groups help their members work collectively to mobilize resources and learn organizational skills, and they either replace government services too remote from the village or help link members to them by reducing transaction costs through scale economies. Clearly, one reason for the appeal of these informal associations is the weakness of local government. Although such groups would still have a role to play even if local government were stronger, the fact that it is not increases the demands on such groups. Of course, while they are able to perform certain important functions, these groups are no substitute for local government. While they can resolve disputes within a group, for example, they have no authority and little influence beyond their own membership. And while they can mobilize meager develop resources of their own members, they have little or no influence on the allocation of government funds earmarked for village development. Clearly, a strong, representative local government system is needed to combine the grassroots appeal of informal groups with the broader powers and authority of government.

Specific Institutional Issues

A few specific institutional concerns affecting agricultural development deserve special mention.

Table 9.3 Comparison of state-wise short-term credit requirement and credit supply for crop production in 1984-85

Sr. No.	States	Supply Credit	Requirements Credit	Percent of Credit Supply to Requirement
1.	Jammu & Kashmir	5	230	2
2.	Himachal Pradesh	5	131	4
3.	West Bengal	101	2,959	3
4.	Assam	3	870	*
5.	Punjab	351	2,118	17
6.	Uttar Pradesh	281	3,252	9
7.	Bihar	43	1,830	2
8.	Orissa	87	2,172	4
9.	Andhra Pradesh	493	2,339	21
10.	Haryana	187	1,284	14
11.	Rajasthan	119	1,048	11
12.	Gujarat	223	1,849	12
13.	Madhya Pradesh	185	2,236	8
14.	Maharashtra	352	2,776	13
15.	Karnataka	259	1,466	18
16.	Kerala	429	419	102
17.	Tamil Nadu	367	2,298	16
	Total	3,490**	29,277	12

* Less than one percent

** Short-term credit not adjusted to the concept discussed earlier

Source: Desai, D.K. (1988)

Credit is well known to play an important role in facilitating investment in improved agricultural technology. The positive production elasticities for banks in table 9.2 provide evidence in this regard. Likewise, table 9.3 (from Desai, 1988) shows that, on the whole, those states with better agricultural performance have a higher volume of short term production credit per ha.

In most of India, weak formal banking and cooperative systems provide subsidized credit, but defaults are extremely high and funds are provided disproportionately to relatively large farmers. In addition, occasional interference by politicians to forgive farmers' debts only serve to weaken the banking system. Desai (1988) estimates that, excluding Kerala, the ratio of credit supply to farmers' short term credit requirements in India is about 1:10. Meanwhile, informal village moneylenders provide coverage to a wider range of clients but at very high rates of interest. Hanumantha Rao and Gulati (1994) indicate that many village moneylenders borrow from the formal sector at concessional rates in order to relend to their poorer neighbors at higher rates.

Hanumantha Rao and Gulati suggest that for most farmers, the advantages of subsidized interest rates offered by the formal sector are far outweighed by the fact that formal sector funds often are not available due to rationing and bureaucratic hassles.

They suggest that the neediest farmers would be made better off if concessional lending were abandoned and bank managers were given more autonomy and protection against political interference. Banking operations could be made simpler and more decentralized in order to reduce transactions costs of both banks and their clients. Higher interest rates would help banks become viable credit institutions rather than merely a means for channeling concessional funds. Under these circumstances, banks could attract deposits, and they would have more incentive to develop better loan portfolios.

In short, this step would help develop greater professionalism in the banking sector (Hanumantha Rao and Gulati 1994). Land tenure is another issue in which some reform is needed. As mentioned in the section on natural resource management, land-to-the-tiller laws have led to a situation in which tenancy is widespread but unofficial, and leases are limited to a year or two in order to avoid potential ownership claims by tenants. From a natural resource conservation perspective, this system makes it likely that tenants' management decisions will be guided by short time horizons.

Not surprisingly, Pender and Kerr (1996) have found in a village level econometric study that land under tenancy is less likely to receive soil conservation investments. This situation might not hold if longer term leases were permitted. Even aside from natural resource management concerns, freeing the tenancy market would make it easier for landowners to allocate land to its most productive use. Jodha (1984) shows that much of the lease market in semi-arid areas comprises plots leased by smaller farmers to larger farmers.

An active lease market enables such farmers, who may lack the means to cultivate their land in any given year, to earn income from it yet retain full ownership. In other circumstances, land owned in larger holdings is often kept fallow by absentees who have lost interest in farming but wish to retain their land as a long term asset (Kerr and Sanghi 1992). An active lease market would encourage such farmers to lease their land to be used productively rather than left fallow. Land taxes, meanwhile, would reduce the value of absentee land holdings and possibly encourage sales by absentees.

PRICE AND TRADE POLICIES

Indian agriculture is subject to a wide range of policies, including export restrictions ranging from licensing requirements to complete bans for certain products, input and output price controls, and interstate trade restrictions. We do not provide a detailed description of economic policies affecting agriculture here.

Recent years have seen the realization that despite heavy subsidies for inputs such as fertilizers, credit and irrigation, most of Indian agriculture is net taxed, not net subsidized (Gulati et al 1989). Taxation comes in the form of direct restrictions on interstate and international trade, output price controls, and an overvalued currency that discriminates against tradable goods sectors such as agriculture. Within the agricultural sector, policies have been mixed in their impact on irrigated and rainfed agriculture. Some policies are crop-specific; for example, oilseeds have been heavily protected in recent years, while cotton has been discriminated against (Gulati et al, 1989). Both have large areas under rainfed conditions.

In other respects, however, irrigated agriculture clearly has been favored. We have already mentioned the infrastructure bias in favor of irrigated areas and its implications for agricultural development. In addition, canal irrigation is heavily subsidized; not only do users not bear the massive investment costs, but the charges they pay are so low they cannot even cover maintenance costs. Well irrigation investments are almost entirely borne by users, but operations for most users are heavily subsidized via underpriced electricity to power pumps. Gulati et al (1989) found that, thanks to subsidies on both canal and tubewell irrigation, farmers in Haryana and Punjab (the two most productive agricultural states) faced more favorable input/output price structures than in other states.

Economic reforms initiated in 1991 will open the Indian economy to greater international exposure, reducing protection of industry and reducing taxation of agriculture. Higher prices of agricultural commodities are expected to provide incentives for increased agricultural output. However, as Hanumantha Rao and Gulati (1994) point out, to have the desired effects on producers' incentives, various domestic marketing reforms will be needed, such as removal of interstate trade restrictions and state procurement monopsonies. One drawback of liberalization mentioned by Hanumantha Rao and Gulati is the reduction in public funds available for research and infrastructural development. They argue that public and private infrastructural development are complements, a position supported by the evidcnce of Subbarao (1985) and Binswanger et al (1993) presented earlier in this section. Another possible drawback of liberalization would arise in the event of high prices of foods that act as wage goods for poor rural people. Parikh et al (1995) and Hanumantha Rao and Gulati (1994) both argue that employment programs and targeted food subsidy programs should be retained in order to protect poor people.

Parikh et al (1995) simulate some of the effects of liberalization using a computable general equilibrium model. Some of their conclusions are as follows:

- Trade liberalization stimulates economic growth by increasing real investment in agriculture due to improved terms of trade and by increasing allocative efficiency, both within agriculture and between agriculture and other sectors. They find that the trade liberalization impact is greater than the allocative efficiency impact, implying that nonagricultural trade liberalization is more important for Indian agriculture than agricultural trade liberalization. This is because nonagricultural trade liberalizations alone will help steer investment funds toward agriculture.

- They find that liberalization could stimulate Indian exports of several crops, including rice. Due to the thinness of the international rice market, however, Parikh et al advise that rice export tariffs would be needed to limit rice exports in order to prevent its price from dropping precipitously.
- They find mixed implications of removing agricultural input subsidies. If this measure is taken alone, it would aggravate rural poverty while benefiting the urban population through lower taxes. On the other hand, if funds currently earmarked for input subsidies are instead directed toward further irrigation development, the rural population would enjoy net benefits, and these benefits would be more equitably distributed than current input subsidies.

10

SUMMATIVE ANALYSIS

India's rainfed agricultural sector provides livelihoods for hundreds of millions of people, and it is the source of nearly half of the value of the country's agricultural production.

As unexploited irrigation potential is increasingly scarce, planners look increasingly to rainfed agriculture to contribute to food production and economic development in the decades ahead. The material presented in this paper has shown three main points:

(1) irrigated agriculture has always been more productive than rainfed agriculture, and it probably always will be;

(2) several types of rainfed agriculture have been highly productive, particularly in the last decade, thus providing hope that the rainfed sector can in fact make major contribution in coming years;

(3) there are numerous constraints facing rainfed agriculture, and numerous possible approaches to overcoming them. In this section we review briefly some of the issues involved in rainfed agricultural development, and outline strategies for supporting agricultural development to achieve broadly defined goals of productivity, equity, and environmental sustainability.

DISTRICT LEVEL DATA ANALYSIS

A significant portion of this paper is devoted to a district-level analysis of the sources of productivity in rainfed agriculture. The two parts of the analysis are the estimation of a production function to identify the relative contributions of different factors, and a tabular analysis to examine district level agricultural growth rates by agroecological zone and irrigation status.

The production function analysis yields expected results on the whole; it shows the important positive contribution of relatively high rainfall and irrigated area, for example. Some other findings are somewhat unexpected; for example, tractors show a more positive contribution than expected, markets show a significantly negative effect, and literacy rate shows a negative but insignificant effect.

The production function analysis demonstrates a useful approach to analyzing the sources of agricultural productivity, but the analysis presented here most likely suffers from insufficiently detailed data. For example, some infrastructural variables are available but others, such as electrification or credit and input markets, are not. This could cause omitted variable bias, meaning that the coefficients of certain variables will reflect the effect of other missing variables that are correlated with them.

For example, the high, positive coefficient of the tractor variable could result in part from other factors associated with tractor use, such as a strong off-farm economy or better infrastructure. The negative effect of markets on output indicates the need to conduct such analysis in the future using simultaneous equations and more detailed data in order to capture possible indirect effects of markets on output that may have been missed here. Similarly, dummy variables representing each agroecological zone may capture a wide variety of information, but we do not have the means to disaggregate it. This means that it is critical to expand the set of variables in the district level data set in order to obtain more conclusive results.

Another limitation of the district level analysis is the absence of good data on performance indicators of factors other than productivity, such as poverty and natural resource degradation. In the paper, lack of such data limited us to a review of existing literature, which of course suffers from the same data shortage. With a complete set of data, one could analyze district characteristics that determine poverty levels or natural resource conditions or their changes over time. A critical issue here concerns the appropriate indicators for such analysis; poverty, for example, can be represented by market wage rates or by the percentage of people under the poverty line. Any indicator will have its limitations, of course. Environmental indicators may pose greater difficulties; possible candidates include forest cover or its change over time, estimates of soil degradation and its changes over time, etc. The tremendous diversity of natural resources over small areas may suggest that the district is actually too large a unit for such analysis, because different parts of a district may face different kinds of problems. Researchers at ICRISAT and the Indo-Swiss Livestock Project have recently used mandal level data to analyze crop and livestock production patterns in Andhra Pradesh; this may prove to be a more appropriate level of analysis for certain problems. As in all socioeconomic research, the utility of using the more disaggregated data depends on the additional costs of collecting it and the additional benefits of the information that it is expected to yield.

The tabular analysis was designed to provide a disaggregated analysis of growth rates over time under different conditions. Other studies have conducted the same analysis on an aggregated scale, for example to compare agricultural growth in predominantly irrigated vs predominantly rainfed districts, or to compare agricultural growth before and after the green revolution. Our analysis also does this, and it agrees with other studies that growth rates have been roughly constant over the entire period. Our study further attempts a more disaggregated analysis in order to identify the sources of growth in different periods. This approach can help us examine, for example, whether irrigated wheat regions drove growth in the first 10-15 years of the green revolution but favorable rainfed areas drove growth subsequently. Our study also disaggregates rainfed agriculture into different types, unlike other studies. This enables us to identify variations in the performance of rainfed agriculture on the basis of region and other characteristics.

Our analysis does not show conclusive results regarding changes in the sources of growth over time. One reason for this may be that in tabular analysis, the data cannot be disaggregated perfectly, and somewhat arbitrary decisions are required to classify districts by irrigation status and to delineate time periods. More persistent efforts to analyze the data under alternate specifications of time periods and irrigation status may or may not result in more striking findings.

Our analysis also does not show conclusive evidence regarding the determinants of growth for different rainfed agricultural types. The tabular analysis shows that rainfed agriculture has grown slowest in the green revolution areas of the northwest, where irrigated agriculture has driven large increases in overall food production. Rainfed agriculture grew at less than 1% during the period 1968-84, and less than 2% in the period 1984-91. In the rest of the country, rainfed agricultural production grew at rates of 1.4% to 3.5% percent during the period 1968-84, and 2.9% to 4.7% between 1984 and 1990 (table 4.1). The production function analysis is designed to help explain the factors that drive these variations. It shows that rainfall levels are perhaps the most important determinant; it also supports the point made by earlier researchers that differences in the level of infrastructure (particularly roads) help determine rainfed agricultural growth rates. However, the data are insufficiently detailed to provide a more definitive picture of the causes of variations in growth rates for different rainfed agricultural types. Subsequent analysis will address this question.

SPECIFIC ISSUES

Agricultural Research and Extension

Evidence suggests that public and private sector agricultural research efforts have been highly successful in developing seed technology that is widely adopted and highly productive in irrigated areas and favorable rainfed zones. The strong performance in recent years of rainfed rice in eastern India and rainfed sorghum in central India provide the basis for optimism that rainfed agriculture can in fact be an important source of agricultural production in the coming decades.

For soil and water management technology in these areas and for all kinds of agricultural technology in unfavorable rainfed areas, however, agricultural research has had limited impact. Evidence suggests that the current approach of developing technology packages on research stations and then transferring them to farmers' fields has had limited effectiveness. This is the case for two reasons. First, the diversity of rainfed agricultural systems may require location specific approaches to soil and water management, so that a single system developed in isolation on a research station may not be widely applicable. Second, farming systems in marginal areas are highly diversified, and they coexist with other nonagricultural activities that comprise a household's livelihood strategy. As a result, new land management technologies developed on research stations may interfere with other components of existing farming systems. If this is so, then adopting the new technologies imposes opportunity costs on farmers who must adapt or sacrifice other components of their existing farming systems, which of course makes the new technologies less attractive. This implies the need for more diagnostic work to understand farming systems better. It also suggests that developing new technology in participation with farmers, building on their existing farming systems to improve soil and water management, may yield new technologies that are both effective and widely adopted.

The extension system faces a similar set of challenges. The extension system traditionally has followed a one-way system of communication; it transfers technology developed on research stations to farmers. There is no formal system for reversing the flow of communication, so scientists are rarely in direct correspondence from farmers in order to receive specific, detailed reactions to the new technologies developed.

The extension system may become more effective if it serves as a channel for more effective two-way communication between farmers and researchers. Similarly, the extension system can benefit from increased farmer-to-farmer extension.

This is the case for two reasons. First, farmers are likely to understand each other's objectives and constraints better than outsiders, so they can communicate more effectively. Second, the large discrepancies in agricultural productivity within villages may suggest that there may be significant scope for transferring knowledge from more productive farmers to their neighbors. Much of the difference may be attributable to differences in soil types and other constraints, but some evidence suggests that differences in technical knowledge also matter.

Irrigation

One obvious approach to developing agriculture in rainfed areas is to expand the area under irrigation. Irrigation is the agricultural investment of choice among private farmers in semi-arid areas (Pender 1993), and as we have seen, irrigation enables farmers to achieve higher, more stable yields. Irrigation development, however, can never be more than a partial solution to the problems of agricultural development, because total irrigation potential is only sufficient to irrigate about half of the total cultivated area. The rest will remain rainfed. Nevertheless, it is important to continue to develop new sources of irrigation and improve the economic and technical efficiency of existing irrigation capacity. Equity and environmental sustainability also are critical issues in irrigation development. The major issues are as follows:

- In many water scarce areas, groundwater capacity is nearly fully utilized; additional wells cause water tables to fall, in some cases depleting aquifers or leading to saltwater intrusion. In large part, this problem results from the low, flat rate charged for electricity to power wells and the lack of property rights assigned to groundwater. Serious efforts are needed in water scarce areas to link pumping charges to the volume pumped, and to develop effective property rights to groundwater. Little effort has been made in reforming power prices and water property rights, but there is growing awareness of the problems and some movement toward developing solutions.
- The status quo in well irrigation in dry areas potentially is highly inequitable, since it favors farmers who can afford to continually deepen existing wells or dig new ones. Water markets are a mitigating factor that enable farmers without wells to enjoy the

benefits of groundwater. Groundwater markets have developed rapidly in some areas but slowly in others, and their competitiveness varies as well. Shah (1993) has shown that water buyers receive the most favorable prices when electricity is charged at a flat rate, but in dry areas this pricing system leads to overexploitation of the resource (Kerr et al 1996). Additional work is needed to see how to further develop water markets in dry areas in ways that are both equitable and environmentally sustainable.

- Protective or supplementary irrigation of dryland crops is potentially a powerful mechanism to spread the benefits of irrigation over a much larger area and to increasing numbers of farmers (Dhawan 1988b). Many dryland crops show substantial yield increases resulting from one or two protective irrigations, yet in many water-scarce regions, irrigation water is used intensively for such crops as paddy, sugarcane and horticultural crops, while dryland crops remain purely rainfed. Further work is needed to understand the private and social costs and benefits of extensive vs intensive irrigation, the circumstances under which farmers practice one as opposed to the other, and policy tools that can be taken to encourage the most efficient use of irrigation water.

- Groundwater irrigation is less developed in many more favorable rainfed areas, such as eastern India. This is partly due to the fact that water is less limiting to crop production in such areas, but also possibly to complex tenure relations that inhibit long term land improvement investments such as wells (Repetto 1994). Irrigation development would increase dry season cultivation in these areas, with potentially strong implications for increased production. More work is needed to assess private and social net benefits of well investment in these areas and constraints to socially optimal investment.

- In canal irrigated areas, the area that actually receives irrigation water can be increased through better management of canals that leads to a greater transfer of water from "front end" to "tail end" water users. Engineering and social organization solutions can be combined to organize water users into smaller, more cohesive groups in order to facilitate more efficient and equitable distribution of irrigation water.

SUSTAINABLE USE OF FRAGILE LANDS

Much of the effort devoted to increased productivity of rainfed agriculture revolves around land use planning for integrated use of different types of land. Watershed development is the vehicle by which improved land use principles are promoted. Several problems plague the concept of improved land use, and various steps are needed to solve them.

- According to ICAR, large areas of land are used in ways that are inconsistent with their capability; for example, sloped plots with shallow soil are used to grow field crops even though perennial vegetation is recommended to reduce soil erosion and build up soil nutrients. While unsustainable land use is undoubtedly widespread, it is worth studying indigenous farming systems to make sure we understand them well enough to know their sustainability implications. In many cases farmers do undertake steps to protect resources, but outsiders do not recognize them as such because they do not resemble practices developed by scientists (Kerr and Sanghi 1992). Often understanding a natural resource management system requires observing it in a dynamic context, since what we see in a single point in time may be misleading. For example, many farmers in hilly areas deliberately induce erosion within their plots in order to encourage natural terracing. A casual observer visiting the site in its erosion stage might miss the point.

- Similarly, where farmers are using land unsustainably, there is a need to understand better why they do so. Understanding the determinants of land use is needed to identify the constraints to change, and thus to developing effective, adoptable alternatives. For example, often the argument is made that farmers use environmentally damaging practices because they do not know any better, or because they are too poor to be concerned about sustaining future productive capacity. On the other hand, a growing body of evidence suggests that farmers do perceive environmental degradation and know how to reduce or prevent it, and that inappropriate policies and institutions may be more to blame than ignorance or poverty in leading to degradation. Examples in the Indian context include soil erosion and the adoption of measures to prevent it, and the management of various common property resources.
- Watershed management projects provide a useful context for identifying the conditions under which farmers are or are not willing to adopt recommended practices. Technology transfer in most watershed projects has been so heavily subsidized, however, that it is extremely difficult to learn whether farmers accept technology for its own sake or simply to obtain subsidy benefits. Watershed projects also present a means for experimenting with alternate approaches to technology development and social organization leading to more efficient and sustainable land use.

But most projects are managed so inflexibly that inhabitants have little say regarding the design and implementation of project interventions. They are also heavily oriented toward promotion of new technology. Yet evidence from around the world shows that cases of sustainable rural resource management have relied not on externally introduced technology but rather sound economic policies, infrastructure and institutions that give villagers the incentive to manage natural resources better and encourage them to put their existing knowledge to better use. Watershed projects so far have been an unexploited opportunity for developing innovative approaches to land use, both at the individual and community levels.

- The district level analysis was not able to address issues related to environmental degradation because the data were inadequate for the task. Environmental indicators have two kinds of shortcomings. First, rarely are they available on a district level basis, and second, many are not available on a historical basis, making it difficult to relate their changes over time to possible causal factors. Soil degradation is a good example; the NBSS&LUP in Nagpur has recently mapped the erosion status in various regions of the country, but the data are not matched to districts. With some effort this could be done on a very rough basis, but only for one period in time (the early 1990s). Some data, such as changes in approximate area covered by forests or percent of groundwater utilized, could be constructed for at least some part of the period under study. Additional efforts to construct such data may be useful. On the other hand, in some cases diversity of natural resource condition might justify the use of a smaller scale of analysis, such as the taluk or mandal. Likewise, under this approach analysis could be undertaken even if data are available only from a few states.

District- or even mandal-level data can be critical for analyzing some determinants of natural resource management, such as different institutional or legal approaches or the effects of pricing and marketing policies. However, village and household studies also are needed to understand people's knowledge about natural resource management and mechanisms by which they make decisions. A balance between more aggregated studies that yield "big-picture" trends and village-level studies that yield the details of how decisions are made is needed to understand what drives environmental degradation and devise policies and institutions needed to reverse it.

Infrastructure

The few studies available on infrastructure suggest that it has a strong positive impact on agricultural development, and that government infrastructure investments are concentrated in more favorable agroclimatic areas. The analysis in this study showed little if any effect of roads and regulated markets, the two infrastructural variables available, on agricultural output. However, this may be due to insufficient disaggregation of the data. Infrastructure, like environment, is a subject on which additional district level variables are needed to facilitate more detailed analysis of its contribution to agricultural development. In the present analysis, the data on kilometers of roads and the number of regulated markets in a district may (or may not) be correlated with other types of infrastructure such as access to credit, specific inputs, or public transportation, for example. In any event, it would be beneficial to conduct the district level analysis with more detailed infrastructural data that captures 1) more types of infrastructure and 2) its quality.

Additional types of infrastructure might include health and educational facilities, formal and informal credit sources, electrification, and extent of public transportation, to name a few. More importantly, the quality of infrastructure is critical to whether it stimulates the economy. Formal credit services, for example, are well known to ration credit, so a variable such as the number of banks may indicate nothing about poor people's access to credit. The number of markets, likewise, may not indicate whether agricultural inputs are available to all who need them on a timely basis. All types of infrastructure are subject to similar questions about quality — do schools attract students or teach them anything? Does the current run in electrified villages? Indicators of the quality of infrastructure may be difficult to obtain, in which case proxies may be sought. The presence of the Integrated Rural Development Program (IRDP) or other such schemes may give an indication of widespread access to credit or inputs, for example. Perhaps the best indicator of access to credit is the prevailing market rate of interest from informal sources, but this information would be available only through village-level data. Quality indicators for other kinds infrastructure, such as delivery of electricity or petroleum-based energy, perhaps may be obtained from appropriate state government offices.

Decentralization and Local Institutional Development

In many respects, decentralization and local level institutional development are the "hot item" expected to overcome numerous constraints to agricultural and rural development. Both theory and experience support arguments in favor of rural infrastructural development, but it is too early to know what will happen as India embarks on the Panchayat Raj decentralization plan. The concept behind decentralization is that local governments will be more aware of local problems and peoples' priorities, and that they will have more at stake in delivering high quality service. State government planners, on the other hand, cannot be expected to have as clear an understanding of local concerns, nor are they as accountable to specific constituents in remote areas. Evidence in favor of this view is limited, coming mainly in the form of 1) the positive experience of strong panchayat government in West Bengal and 2) the rise of informal local organizations, such as thrift groups and natural resource users' groups, that have helped their members raise funds or promote collective action, among other things.

Some observers believe that the local groups' informality is their greatest strength. By remaining small, with a limited agenda, they can focus on issues of specific interest to the group, which is defined by the homogeneity and common interests of its members. These groups may select their own members and are accountable to and controlled by only themselves, which simplifies their operation. Neither government bureaucracy not local elites can do much to obstruct them. For these reasons, a formal local government system may not be able to duplicate their strengths.

Others, meanwhile, suggest that the logical step to follow the positive experience of informal local organizations is to develop formal, institutionalized local organizations with authority extending beyond the group. The panchayat would embody this approach. Proponents anticipate that it will combine the advantages of local participation and organization with access to government resources and real authority over how they are allocated. Of course, even if panchayats cannot duplicate all the advantages of informal organizations, there is no reason the two cannot coexist. In fact, informal groups may become more effective under the Panchayat Raj system because they can more easily influence a government that operates at the local level as opposed to the state capital. Whether the Panchayat Raj will be able to stimulate rural development will depend ultimately on the quality of governance by panchayat bodies. Will local government be truly representative and address development objectives of a broad spectrum of the rural population, or will it be dominated by local elites, so that they be in a better position than ever to channel funds and other resources in their favor? The semi-feudalistic history of rural India makes it easy to envision the latter scenario. Information is probably available on determinants of the quality of local governance, but it was not available for this study. In any event, some steps will be needed to nurse the panchayat system to become a healthy, mature democratic institution.

It is important to point out that the Panchayat Raj and other institutional innovations, such as Joint Forest Management, indicate that the central government is committed to diffusion of authority to the local level. This process can only be expected to move gradually, but evidence suggests that it is certainly moving. In time, many other institutional innovations may follow from the emergence of the Panchayat Raj; for example, it may spawn new mechanisms for managing common property natural resources such as groundwater.

Other Issues

Price policy reforms are underway and are likely to have a favorable effect on agriculture relative to other sectors. The limited evidence is mixed regarding the likely relative impact on rainfed and dryland crops; the effect will depend largely on the crop in question. Oilseeds, which are grown widely under rainfed conditions, are protected and will become less favorable under price reforms, while cotton has been taxed will become more favorable with reforms. Additional research currently underway at ICRISAT and NCAER is expected to provide information about the likely effects of reforms on other crops. Technical assistance, information dissemination, and income support measures may be needed to help farmers in the transition from crops that become less favorable to others that become more favorable.

Production risk. Although the unreliable weather in many rainfed areas causes growing conditions and yields to vary greatly across years, farmers have developed various coping strategies to insulate themselves from income risk, at least to a certain degree. As a result, even if individual crop yields vary greatly across years, farmers' incomes may not, so increased yield variability of HYVs is not necessarily a deterrent to adoption. Moreover, research on improved pearl millet seeds shows that high yielding hybrids will often be superior to traditional varieties even under poor rainfall conditions, implying that higher expected yield does not necessarily translate to greater yield risk.

Farmers' drought management strategies fail, however, in the event that widespread drought causes crop failure over a wide area and depresses the rural economy so much that all sources of income are affected. Such aggregate level, covariate risk calls for government intervention to help stabilize incomes and prevent famine. Rural employment and food subsidy programs deserve credit for reducing drought-related hunger in India in the last two decades. On the other hand, government-sponsored rainfall insurance schemes have probably not contributed to increased adoption of improved seeds, but they have done a great deal to drain public funds.

Poverty alleviation programs can play an important role in supporting the poorest rural people, and they have the potential to help mitigate the effects of price reforms on those people whose incomes will fall with reforms. Poverty alleviation programs enjoy widespread political support, but they face two challenges: 1) to operate in as cost-effective a manner as possible, and 2) to stimulate the creation of long term development assets. Many observers argue that employment programs represent the best way to achieve these objectives while also alleviating poverty, but only under certain conditions.

Most importantly, they should offer wages slightly below the prevailing market wage in order to attract people who really need assistance and to minimize distortions in the rural economy. Second, using employment programs to create long term assets is more complicated than it first appears; in many cases only illusory assets are created (Jackson 1982; Kerr et al 1994). As a result, such programs should be developed only after experimentation on a small scale yields an understanding of their ability to create long term development assets.

Human capital development is often mentioned as a critical step toward stimulating the rural economy. Educated farmers can more easily process information and thus are prepared to make better decisions; they also may contribute to the diversification of the local economy, because they will have more to offer to a variety of nonfarm economic activities. Walker and Ryan (1990) show that household level education is negatively correlated with poverty; Hazell and Singh 1993 find the same. Also, better educated farmers may impose pressure on government to be more accountable to its constituents and serve them more effectively.

The production function analysis in this study provided counterintuitive results regarding the contribution of literacy to the value of agricultural output. The reason for the negative relationship found between the two is not clear; perhaps it is simply an anomaly of the data resulting from the fact that many infrastructural and institutional variables are not available.

Despite the findings of our analysis, it seems reasonable to argue in favor of increased investment in education in rural areas. Education will contribute to economic diversification, which probably represents the long term solution to development of less favorable areas, and it will contribute to declining rates of growth in the population.

Tradeoffs Between Investments in Different Types of Agriculture

One of the questions motivating this study concerned the likely returns and tradeoffs involved in diverting development resources from more favorable to less favorable areas. Unfortunately the quantitative analysis did not shed much light on this subject, but the review of existing literature offers some useful insights.

Binswanger et al (1993) provide evidence that more favored areas receive more infrastructure investment, which in turn creates further regional disparities. Their finding of the positive impact of infrastructure may suggest that areas with less developed infrastructure will have higher marginal returns to additional investment, in which case shifting resource allocation in favor of less developed areas would be more efficient as well as more equitable. More decentralized infrastructural investment allocation under the Panchayat Raj may lead to less biased allocation between regions, as long as funds are distributed equitably among panchayats.

Other recommendations listed above argue for changes in the way problems are addressed, but they may not have major implications for interregional or intersectoral resource allocation. Agricultural research in marginal areas provides a good example. As shown in section 4, there is no bias against marginal areas in the allocation of resources for agricultural research, and no strong argument in favor of shifting resources from favorable to marginal areas. Instead, the main argument presented here regarding the allocation of research resources in marginal areas is that there should be a shift in emphasis toward on-farm research and toward solutions based increasingly on social organization, not just technology. The shift in resource allocation that this would imply is strictly within the region and the sector, not between sectors. The key challenge is to change the culture of agricultural research to overcome the aversion to working in farmers' fields and the perception that farmers have little to offer the research process. This challenge is by no means trivial. Researchers need stronger incentives to conduct more on-farm work. In many cases they will need help in building collaborative arrangements with people, such as extension workers or NGO officials, for example, who can help bring them in contact with farmers.

As mentioned above, there is a distinct trend toward the view that the solutions to rural problems lie with rural people. The introduction of Joint Forest Management and the Panchayat Raj provide strong evidence of this shift. Meanwhile, gradual movement is taking place toward more participatory approaches to watershed management, technology development, and other activities. Truly participatory approaches are still in the minority, but they are less frequently treated as anomalies. Government projects are increasingly attempting to become participatory in nature, and although their progress is slow, it is also steady.

Steps toward Increased Participation of Rural People

As with most problems, it is easier to diagnose shortcomings than to prescribe solutions. Agricultural development is prone to fads, in which recommendations come in style with little solid evidence of their worth and then eventually go out of style. In this paper the main strategy argued for is more decentralization and participation in planning and developing infrastructure, institutions, and technology. While there is growing evidence of the advantages of participatory approaches to research and development, it remains fairly scattered and anecdotal. In India, most of the evidence comes from the voluntary sector, and much of it is not well documented.

Agricultural researchers and development project managers in the government sector do not have ready access to this material, and there is little incentive for them to try to obtain it. In recent years participation has become a "buzz word" in development circles, and many government projects have paid lip service to the term. However, they are not truly participatory because local people still have little or no influence in project planning and implementation. Meanwhile, government officials exposed to these projects may rightly argue that the so-called participatory approach has not yielded any particular benefits, and participation will have a bad name. This is the surest way to turn participatory development into a passing fad.

Efforts are needed to expose more scientists, planners and project managers to truly participatory methods and train them in their use. This should be done incrementally, for several reasons.

First, arguments in favor of greater participation are based on relatively limited experience, and they require further testing in the field before embarking on them in full force. Special grants may be made available to researchers interested in pursuing on-farm research, or to project planners who wish to try new approaches on a small scale. It is important to note that ICAR tends to promote research that follows fairly strict methodological guidelines which, incidentally, do not favor participatory research. Various government ministries organize development projects in the same manner: large scale programs follow strict implementation guidelines, with little flexibility for experimenting with project design.

The resulting "all-India coordinated project" approach inhibits testing ideas that fall outside the realm of the specific guidelines. Moreover, such a rigid approach means that when project guidelines change, they apply to everyone. Radical changes in project guidelines would be highly risky, in case the new approach does not work. A better approach would be to build flexibility in project design, so that some special, small scale projects are allowed to try new approaches.

Researchers and project managers working under such circumstances would have to adapt to a variety of local conditions, and they would have to be willing and able to accept and learn from feedback from farmers. A second reason why participatory approaches should be implemented gradually is that researchers and development planners cannot be transformed overnight. People have spent entire careers in these fields without ever seriously seeking input from local people, so change will come gradually. Special grants for participatory projects would attract creative, field-oriented researchers and project managers. Their more conservative colleagues could observe their work and be influenced accordingly. A third, important point is that some problems are more suited to participatory approaches than others. In agricultural research, for example, traditional research-station approaches are highly suited to activities such as mapping plant genes or analyzing soil chemical and physical processes, for example. The argument in favor of increased collaboration between scientists and farmers is not an effort to dismantle the existing research system. Rather, it is an effort to change the culture of research so that scientists understand problems from farmers' perspectives and design technologies that are more applicable to farmers' conditions.

Priority of Agricultural Development

Geographically villages take a lion's share in India. More than before the priority now is for the agricultural sector and without agricultural development India cannot firmly achieve double digit growth rate in economy. It is doubtful that the failed political parties of the past in power now with the corrupt government agencies implementing the economic programs can deliver the expectations of the mass in the rural India. However, the aspirations of the people can be fulfilled only if the corrupt officials are removed and the implementation of economic reforms is done with careful monitoring mechanism.

AGRICULTURE SECTOR

All the nations of the world give first preference to Agriculture before other sectors. But in India importance for development is given in the reverse order of preference. By the persistent efforts of the peasants depending only on the monsoons and a little sop once in every five years India has achieved surplus production of food grains in the 1980s. Also, annually though India is capable of producing 220 million tons of food grains in surplus, because of poor distribution system, it has become incapable of preventing the farmers from deaths due to starvation or committing suicides due to debt burden, long drought situation or floods. Further, because India has a storage facility of 70 million tons only for keeping food grains in stock and the perishable nature of vegetables and fruits, 35% of agro-products are lost as waste. So, it is a pity that India suffers from the problem of scarcity in plenty because of bad policies and poor infrastructure facilities.

FOODGRAIN PRODUCTION

As all funds are allotted to industries, defense and commerce out of the 30% revenue available from the national income after paying 70% of it as salaries to the government employees, farm sector is neglected without any reforms and modernization process being carried away by the notion that agriculture is a loss making sector. So, with no hope of economic development in rural areas many people have migrated to metropolitan cities and towns to work in industries, mills and companies. Satellite towns have increased along with the population. So, to balance the increase of population and their needs food grains production has to be increased to 300 million tons before 2020.

LIBERALIZATION OF ECONOMY

Consequently, for mobilization of financial resources to providing funds to various development projects, liberalization of economy has been started some 15 years ago leading to privatization of public sector units (PSU) and globalization of trade. But development of industries, IT, textiles, leather and other sectors at the expense of agriculture with the view to get short term gains at the cost of long term consequences has led to the present chaotic crisis in the 80% of rural India compelling the government to fulfill their demands on war footing by hook or crook.

AGRO-INDUSTRIES

So, infrastructure in the rural areas has to be developed in India first by introducing food for work program that will pave the way for laying of roads, digging of wells, installation of electric power, construction of go downs with refrigeration facilities for storing food grains, processing of vegetables, fruits and flowers into pickles, jams and perfumes for domestic and foreign consumptions and marketing them abroad utilizing export potentialities with the expertise of entrepreneurs. The skill and utilization of entrepreneurs, scientists, engineers, doctors and teachers have to be geared up in rural areas for the spread of education needed for the dissemination of agricultural knowledge, reduction of cost of production and the processing of products using TV, Radio, internet and possible phone facilities and various other jobs. This is the only way to stop migration of villagers to cities.

Above all, with the guidance, monitoring skill and control of sincere scientists, experts and entrepreneurs government, banks and financial institutions should come forward to make financial investments in irrigation projects, for buying of crops, fertilizers and insecticides, reducing cost of production, processing of agro-products and marketing them in the world in order to make agricultural development effectively, beneficially and profitably possible in India. So, priority to agriculture sector is the utmost need of the hour now.

WTO

World trade agreements are binding on all nations in the World Trade Organization (WTO), in which India and some other third world developing countries are members. Only recently many Asian nations by the globalization process have integrated themselves in the world market for the promotion of their trades and development of their economies. Because the advanced western countries are globally well established in trade, they are dominating other countries in the world market.

TRADE/LIVELIHOOD

The success of the developed countries depends upon Market access, Domestic support and Export subsidy that are called as the three pillars of trade. Accordingly, as per their policy of protectionism both the USA and the EU countries give subsidies to their farmers, cut tariffs for exports and impose heavy duties for imports of farm products; and through WTO compel the Asian developing countries to cut subsidies to farm sector and reduce customs tariffs to export of food grains. Agriculture is just a trade for USA and EU but it is not only a trade but also a means of livelihood for millions of farmers in India and Asia.

MEASURES NEEDED

In the Doha WTO meeting India forcefully said that the aspirations of majority of Asian peasants should be taken into consideration for any agreement. Consequently, in the Cancun meeting India along with Mexico, Brazil and South Africa formed into G 20 countries to flight for their demands. But USA and other countries took up the Singapore problems like Investment and competition, Trade facilitation, Procurement of products by government and Regulation of Customs tariff for an inconclusive discussion which were again taken up in the Geneva Conference leading to no definite agreement.

Even after the meetings at Hong Kong and recently at Davos in the World Economic Forum (WEF) also no solid decision has been taken yet. So, the G 20 countries should mobilize majority support in the next WTO conferences to make USA and EU accept the conditions like 1. Ban subsidies to farmers and 2. Reduce Customs tariff for imports of agro-products from other countries which will boost sale of Indian farm products in the world market.

INDIA

Introduction

During the last five and a half decades, Indian agriculture has witnessed numerous changes. The major transformation took place with the introduction of high-yielding varieties of crops in the mid-1960s. This innovation, coupled with investments in irrigation infrastructure, expansion of credit, marketing and processing facilities led to a significant increase in the use of modern inputs. These developments raised agricultural productivity and led to a substantial increase in the supplies of various agricultural products for markets rather than for own consumption by the farm households. The "green revolution" has been followed by the "white revolution", "yellow revolution" and then the "blue revolution", leading to an increase in the output of milk, oilseeds and fish and fish products, respectively.

Agricultural growth, measured as the annual rate of growth in net domestic product, increased from 1.9 percent *per annum* before the Green Revolution period (1950-1951 to 1966-1967) to 2.3 percent during the first phase of the Green Revolution (1966-1967 to 1980-1981). The growth rate accelerated further to 3.1 percent during the second phase of the Green Revolution (1980-1981 to 2000-2001). At the same time, the share of agriculture in total GDP has fallen from about a little over half in the early 1950s to around one-fourth in the late 1990s. Similar changes have occurred in the country's export basket as well. The share of agricultural exports in total exports of the country fell from 44 percent in early 1960s to about 14 percent in the late 1990s.

However, the majority of India's population still continues to depend heavily on the agricultural sector for employment and as a source of income. Seventy-two percent of the population, about 742 million people, still live in rural areas, and their fortunes are also influenced by developments in the agricultural sector to a considerable extent. Coupled with the high incidence of poverty, which is predominantly a rural phenomenon (74 percent of the poor live in rural areas), these facts demonstrate the continuing importance of the agricultural sector in the Indian economy.

The implementation period for the AoA coincided with the reform process which has been under way since 1991 in response to the macroeconomic crisis that occurred as a result of the depletion of foreign currency reserves and a high fiscal deficit. To address this crisis, key reforms were initiated, by way of ending the old industrial licensing regime, more liberal policies towards foreign direct investment, significant reduction in tariffs on industrial goods and currency devaluation.

The focus of this reform process was initially on the manufacturing sector, but gradually these unilateral changes have been extended to the other sectors of the economy including agriculture. In agriculture, although there was no direct government intervention in production and investment decisions of farmers, government did, and still does, influence the legal and economic environment in which farmers and other economic agents operate. Most of these interventions in agriculture were conceived to deal with situations associated with famines and scarcities that existed before the Green Revolution period.

In the case of the domestic market, the withdrawal of the restrictions on the movement of agricultural commodities is one of the major changes that has been brought about during the present reform process. The licensing requirements and stocking limits for the wholesale and retail trade that were a part of the Essential Commodities Act (1955) have been removed recently. The system of Selective Credit Controls, which was used to regulate institutional credit to traders in commodities since 1943, has also been abandoned. Future markets, which were banned since 1942 under various statutory orders and since 1955 under the Forward Contract (Regulation) Act, are also now allowed.

As far as external trade is concerned, in the early 1990s, a few steps to liberalize agricultural trade were initiated about a year and a half after the July 1991 reforms, and these have been followed by a number of prominent reforms thereafter. Now, the exports of all major agricultural commodities, barring a few exceptions such as cotton, onion and niger seeds, have been liberalized. These reforms have brought about some significant changes in agricultural trade, which are discussed in detail in the relevant sections of this chapter.

India is a founding member of GATT. India signed the Marrakesh agreements of the UR as a permanent developing country member of GATT. Its first Trade Policy Review took place in 1993, and its second in 1998.

India's commitments under the AoA

Market access

Commitments on Tariff Bindings

Under market access commitments in the AoA, member countries were required to replace all types of non-tariff barriers with tariffs, and to reduce tariff levels under a time-bound programme. In addition to these commitments, this measure also called for maintaining current access opportunities and establishing minimum access tariff quotas. For countries such as India, where all agricultural imports were covered by QRs for Balance of Payment (BOP) reasons, only ceiling bindings had to be submitted. For these ceiling bindings, there was no upper limit, provided the tariffs had not been bound in earlier rounds of negotiations. In addition, there was no obligation to reduce these ceiling bindings during the implementation period.

India had previously bound only some of its agricultural tariffs. These included commodities such as rice, coarse grains, dairy products and edible oils - rice and dairy products in the Geneva Protocol (1947), maize and millet in the Torquay Protocol (1951), sorghum during the Dillon Round (1962) and soybean and rapeseed oil in the Tokyo Round (1979). For other products for which no tariffs had been bound earlier, India submitted very high ceiling bindings of 100, 150 or 300 percent.

When the QRs were being removed (discussed in detail below), apprehensions were expressed about the likely impact of these liberalization measures on the domestic market, particularly in the case of those commodities for which the bound levels of tariffs were zero percent. These included commodities such as rice, coarse grains and dairy products, which were bound in the earlier rounds of negotiations. To raise bound rates for these products, India initiated negotiations with its trading partners under Article XXVIII of GATT and renegotiated new bound tariffs, which are shown in Table 1.

In the case of dairy products, the new bound tariff rates were 60 percent, and in the case of cereals, the new bindings range between 70 and 80 percent. The other products for which the bound rates were raised include apples (from 40 percent to 50 percent); rape, colza or mustard oil (from 45 percent to 75 percent); and preparations for infant use (from 17.5 percent to 50 percent). In exchange, bound rates of tariffs for some items were lowered. These included items such as vegetables (peas), fruits (oranges, lemons, grapefruit, pears and quinces, prunes), malt, chewing gum, fruits juices (orange juice) and industrial fatty alcohols.

Distribution of Bound Tariffs

As a consequence of these changes, the distribution of final bound tariffs on agricultural products has now undergone minor changes. In agriculture, about 692 tariff lines at the six-digit HS classification level are bound. There are only two items (almonds in shell and shelled) for which the rates of duty are specific in nature; for others, the rates of duties are *ad valorem*. Of the 690 items for which rates of bound duties are *ad valorem*, the proportion with bound duties below 25 percent has fallen slightly, while the number of tariff bindings in the range of 25 percent to 50 percent was increased after the re-negotiations (Table 2). However, there was very little change in the average bound rate. The overall distribution of final bound tariffs shows that about 82 percent of tariff lines have bound rates which range between 75 percent and 150 percent. Approximately 4 percent of tariff lines have bound tariffs of 300 percent.

Distribution of Applied Tariffs

Contrary to these high bound tariffs, the actual applied rates of tariffs on most agricultural products are quite low (Table 3). The distribution of applied tariffs illustrates that, for a little over 89 percent of tariff lines, the applied rates are either below or equal to 50 percent. There are only 9.4 percent of the tariff lines for which the applied rates of duties range between 50 and 100 percent. For only 1.3 percent of items, mainly alcoholic beverages, are the applied rates excessively high, more than 150 percent.

There are only a few agricultural products of significance for which bound rates have become a binding constraint. Among edible oils, soybean oil is one item on which the bound rate of duty is 45 percent. The other items likely to be affected because of the lifting of QRs are alcoholic beverages on which the applied rates of above 150 percent exceed the bound rates on liquor (150 percent).

Changes in Quantitative Restrictions on Imports

As India had maintained QRs for BOP reasons, a few members of the WTO questioned the justification to continue these restrictions with the improvement in the BOP situation in the mid-1990s. Initially, India proposed a time schedule of nine years for the complete elimination of QRs that were maintained owing to BOP reasons. The developing member countries accepted this proposal, but the developed countries were uncomfortable with the time schedule that India had proposed. A group of six developed countries/regions (Australia, Canada, EU, New Zealand, Switzerland and United States) with Japan as a third party initiated dispute settlement proceedings against India. India negotiated a deal with five members with the exception of United States to phase out its QRs over a period of six years beginning in 1997. The United States, however, went ahead and filed a dispute against India. A Dispute Settlement Panel was constituted in November 1997, which ruled against India. India filed an appeal before the Appellate Body, which endorsed the findings of the panel.

As a consequence, an agreement was signed between India and United States under which India agreed to abolish all the remaining QRs that were being maintained because of BOP reasons by April 2001. While these negotiations were taking place with the members of the WTO, India started the process of removing QRs on imports unilaterally. For instance, on the introduction of the eight-to-ten-digit HS in March 1996, imports of 6 161 tariff lines, about 61 percent of the total tariff lines (all imports including agriculture) were already free of QRs. By April 1999, the share of free tariff lines had gone up to around 79 percent. After the agreement between India and the United States, dismantling of the remaining QRs on 2714 items was completed in April 2001. QRs are now maintained on imports of only about 5 percent of tariff lines (538 items) under Articles XX and XXI of GATT on grounds of health, safety and moral conduct.

Implications of the Removal of Quantitative Restrictions on Imports

It is premature to speculate on the exact impact of the removal of the QRs on imports of agricultural products, because the last instalment of QRs was removed only in April 2001. The early trends in imports, however, suggest that the surge in imports that was feared after the removal of the QRs has not occurred. This is indicated by the imports of a few selected items which are being treated as sensitive and whose imports are being monitored on a continuous basis (Table 4).

Table 4. Imports of selected agricultural commodities (US$ million)

S.N.	Product	April-Dec. 2000	April-Dec. 2001	%age change
1.	Milk and milk products	8.9	1.9	-78.7
2.	Fruits and vegetables	287.6	154.6	-46.2
3.	Poultry	Negligible	0.1	n.a.
4.	Tea and coffee	6.1	5.9	-3.3
5.	Spices	28.3	45.3	60.1
6.	Food grains	6.3	0.7	-88.9
7.	Edible oils	1021.5	1 051.1	2.9
8.	Alcoholic beverages	4.8	4.2	-12.5
9.	Cotton and silk	319.3	446.1	39.7

Source: Government of India (2002).

These somewhat early trends in imports establish that the protection provided to Indian agriculture through QRs was really redundant because, in most years, domestic agricultural prices of the major crops were lower than world prices (Gulati and Sharma, 1994; Gulati and Sharma, 1997). Consequently, the abolition of import controls did not lead to large imports of these commodities. This does not mean that the level of imports will continue to be low in the future as well. At the margin, imports of a few selected commodities, which are either not produced or for which the country does not possess a comparative advantage and for which the bound levels of tariffs are low, will certainly increase.

In addition, during the re-negotiations to raise zero bound rates, India had to grant a few concessions, which led to the establishment of tariff rate quotas for five commodities (Table 5). As the in-quota tariffs for these commodities are low, their imports certainly can go up to the agreed limits of tariff rate quotas.

To dispel fears linked to the removal of the QRs on imports, applied import duties on some items have been raised in recent years. Import duties on edible oils, cereals, sugar and, more recently, cotton have been revised upwards. Likewise, import duties on plantation crops, which have witnessed a significant decline in their domestic prices owing to a sudden increase in output and fall in international prices, have also been raised.

To safeguard the interests of the domestic liquor industry, the government has proposed countervailing duties that will be imposed on the basis of levies of state excise that are currently prevalent in different states. This means that the effective rates of import duties will be much higher and will be able to provide protection to the domestic liquor industry for some more time.

Domestic Support

Countries use a number of policy instruments such as support prices and input subsidies, which affect incentives that farmers receive in terms of prices and hence influence resource allocation. In the AoA, the impact of price support and related policies is captured through the AMS. India has a product price support system in the form of minimum support prices announced by the government for different commodities, based on the recommendations of the Commission for Agricultural Costs and Prices. Because official estimates of India's AMS are available for only two time periods - the base period (1986-88) and the first year of the implementation period, which is 1995-1996, we have made our own calculations. This is done mainly to make figures comparable over time as our own calculations show slightly different results because of a few adjustments that we have made. Our analysis shows that for 18 major commodities, the product-specific support, as defined under the AoA during the base period, was (-)US$18.1 billion (Table 6). As a percentage of the value of agricultural output (crop sector), the product-specific AMS is (-)26 percent during this period. In 1995-1996, the estimated productspecific AMS turned out to be (-)34.4 percent of the value of agricultural output, and by 2000-2001, the same stood at about (-)28.6 percent of the value of agricultural output during that year.

The non-product-specific support, which includes subsidies on irrigation, fertilizers, electricity, credit and seeds, was about 1.3 percent of the value of agricultural output during the base period after allowing for exemptions which are granted for resource poor farmers in developing countries. In 1995-1996, the non-product-specific support was roughly 1.9 percent of the value of agricultural output, and by 2000-2001, the same worked out to be about 2.3 percent of the value of agricultural output.

The negative product specific support to Indian agriculture basically shows that various controls on domestic as well external trade have kept domestic prices of major crops below world prices. In the case of domestic trade, these controls included restrictions on the movement of agricultural commodities, compulsory procurement levies, licensing and stocking requirements and credit controls. The controls on external trade comprised export prohibitions, quantitative restrictions, minimum export prices and canalization.

The net result of these policies has been that negative product specific support outweighs the positive non-product specific support. These computations compare external reference prices of the base period (1986-87 to 1988-89) with current support prices. For this reason, these estimates do not show the actual subsidy or the tax which is imposed on the agricultural sector currently. To estimate this, one would have to compare current international prices with current domestic prices, and add the impact of total agricultural and economy-wide policies (Krueger, Schiff and Valdes, 1988).

Export Competition

India does not have a system of direct export subsidies, so it made no commitments on export subsidies. There were a few benefits that were available to the exporters of agricultural commodities through income tax exemptions under section 80 - HHC of the Income Tax Act (1961) on profits from export sales. In 2000, the government decided to phase out these benefits over a period of five years starting from 2000-2001, making profits completely taxable by 2004-2005.

As the agreement allows developing member countries to subsidize costs of marketing agricultural products including handling, upgrading and other processing costs and the costs of domestic and international transport and freight, India is making use of these provisions. The schemes facilitate mainly the exports of horticultural items and are operated by the Agricultural and Processed Food Products Export Development Authority (APEDA).

Because the exports of many agricultural items have been adversely hit owing to the fall in commodity prices and aggressive subsidization by those members which are allowed to subsidize their exports, the government is thinking of extending these subsidies, which are permissible under the agreement, to other agricultural products as well.

Although there are no commitments for India on export subsidies, there are restrictions on introducing direct export subsidies in the future that are not compatible with the agreement. This has become a binding constraint in the light of the huge surplus stocks of cereals which are currently much above the stipulated norms for buffer stocks. Managing these surpluses, which are due to a combination of several factors such as domestic policy, good monsoons and low commodity prices in international markets, has become an important issue.

The significance of this problem can be gauged from the fact that the stocks of cereals have been swelling for the past five years, and currently stand at about 63 million tonnes (May 2002). This level of stocks is about four times the minimum buffer stock norm prescribed for the month of April.

Holding such a high level of stocks is expensive, but the options are limited. Releasing these stocks in the open market will lead to a significant fall in prices, which may benefit consumers in the short run, but will have significant implications for future output growth and food security in the long run. Similarly, exporting at such low prices is also not feasible without resorting to direct subsidies which are not permissible.

Disciplines on Exports

As regards export controls, there were two broad groups of commodities which were treated differently. One group consisted of traditional and newer export products such as tea, coffee, spices, tobacco, oil meals, hides and skins, jute, fish and fruit and vegetables. In the 1960s and early 1970s, exports of these commodities were restricted by export taxes and other controls, but during the late 1970s and the 1980s, policies became more supportive of exports, for example, by removing most export taxes. At the time of the 1991 reforms, a number of complex export regulations to which these products were subject were abolished or simplified.

The other group, which is much larger in terms of production, employment and domestic consumption, includes the main cereals, pulses, sugar, oilseeds and cotton. Exports of all these products have been controlled, in most cases directly by state agencies, and in others, through export prohibitions and minimum export prices. These prohibitions were widespread in the past and even now continue to be an important tool to ensure commodities in the domestic market in case of shortages. The examples include onions, cotton and niger seeds.

For most of these commodities, these controls have now become meaningless in the light of excessive supplies and the removal of QRs on imports. The government constituted a panel of members of parliament, which recommended the abolition of such controls. This recent shift in policy has been notified in the new export-import policy, which was announced in April 2002.

Sanitary and Phytosanitary Standards

The provisions of the SPS Agreement state that "measures taken to protect human, animal or plant life or health, must be based on scientific principles and shall not be applied in a manner that would constitute a disguised trade restrictions." Indian products which have been particularly affected by SPS measures include marine products, groundnut and egg powder. Some cases have also been reported for mango pulp and sugar. In the case of marine products, the EU imposed a comprehensive ban on all fish exports from India in 1997 after some consignments were found to be contaminated with *Salmonella* and *Vibrio cholera* bacteria. The differences in the application of rules regarding such restrictions are evident from the fact that marine products were being exported to the United States throughout the period when there was a ban imposed by the EU.

In the case of groundnuts, aflatoxin was found in shipments of groundnut to the EU. The EU had issued a regulation in 1994 setting down maximum limits to certain contaminants in foodstuffs. In July 1998, the EU added limits for aflatoxin content and also elaborated further the sampling procedures for testing the lots for aflatoxins through a directive. Regarding egg products, a Japanese importer reported the presence of BHC (beta isomer), a contaminant, far in excess of the permissible level which affected exports of egg products.

These cases brought into focus several issues including lack of awareness and the problem of consistency of domestic SPS standards with the standards of the importing countries. To handle the problems originating from these cases, the government has taken several steps. For marine products, the rules for quality control and inspection of exports under the "Export of Fresh, Frozen and Processed Fish and Fishery Products (Quality control, inspection and monitoring) Order and Rules, 1995" have been modelled on the basis of the directives given by the importing countries. These rules also comply with the HACCP quality control methodology. More units are being encouraged to comply with these standards.

The Government of India also set up a National Codex Committee under the Department of Health, Ministry of Health and Family Welfare. The Export Inspection Council and its agencies under the Ministry of Commerce, such as APEDA, have started framing standards for the products and packaging of meat, poultry, dairy and honey, grading and packing standards for spices, walnuts, vegetables, fruits and flowers. The State Departments of Animal Husbandry & Dairying under the respective State Governments have set standards for meat and meat products, poultry and milk and milk products. The commodity boards such as the Spices Board, Tea Board, Coffee Board, Tobacco Board and Cashew Export Promotion Council are independently involved in framing and implementation of standards for their mandated products.

There is no doubt that, ultimately, better SPS standards should lead to the lessening of health risks and should benefit consumers, but the manner in which these standards are being enforced gives an idea of the many problems that are being faced by the exporters of developing countries. Broadly, there are three types of problems. First, there are institutional problems such as what should be the point of inspection and conformity (internal or the point of entry) and who should provide the scientific basis to settle disputes. Second, costs of compliance are becoming highly prohibitive because SPS standards are being changed periodically, which makes it difficult to attain the prescribed norms since the norms are becoming more stringent and are moving targets. As opposed to these periodic changes in SPS measures, the technical assistance to help exporters to match these requirements is simply lacking. Third, regardless of the fact that the agreement encourages multilateral agreements on mutual recognition of equivalence of specified SPS measures, member countries enter into bilateral mutual equivalence agreements. This practice favours imports from some countries over others, which results in discrimination against other members.

Trade-related Intellectual Property Rights

Following the URA, member countries were required to provide protection to plant varieties, by patents, by an effective sui generis system or by a combination thereof. The agreement also called for a review of the provisions of this system, four years after the entry into force of the WTO agreement. In contrast to a more strict patent regime, a sui generis system of plant variety protection is a special form of protection which provides both breeder's exemption as well as farmer's exemption. Many developing countries used to exclude plant varieties from patent protection, but in the case of developed countries, plant breeders' rights are often available since most of these countries are members of UPOV. Plant varieties have so far been protected by UPOV, which provides for recognition of rights of plant breeders.

India did not have its own *sui generis* system for the protection of plant varieties. In 1990, the policy planning committee of the Indian Council of Agricultural Research set up an Expert Group to formulate a suitable sui generis system. This expert group recommended the formulation of a draft Plant Varieties Act, which was completed in 1993. After consulting numerous experts in the field, the government finalized the Draft Plant Varieties Act 1994. The revised draft went through several further revisions to become, in 1995, the Draft Plant Variety Protection Act, which was later modified to become the Protection of Plant Varieties and Farmers' Rights Bill in 1998.

This was introduced in Parliament in 1999 and was finally passed by the Parliament in 2001. It will come into force from the date of notification by the government. This legislation, apart from stimulating appropriate returns on investment for plant breeders, also recognizes the role of farmers as cultivators and conservers, as well as the role of traditional rural and tribal communities in preserving biodiversity, by rewarding them through benefit-sharing and protecting their rights. The bill also has a provision for setting up an authority for Protecting Plant Varieties and Farmers' Rights.

When these changes were being discussed, India faced a number of difficulties arising from the TRIPS Agreement.

These problems mainly originated from the lack of preparation on the part of the government to face the new world order under the TRIPS regime. Not only is India well endowed with biodiversity, but there is a great amount of indigenous knowledge that has been accumulated over the ages. Most of this knowledge has not been documented, and there have been a number of claims to patent preparations, for example of garlic, ginger, coriander and neem. Several plants such as neem have been cultivated in the country for their medicinal value, and products of these plants have been used in either primary or semi-processed forms. A European company, W R Grace and Co, obtained a patent for the "anti-fungal properties" of neem. The company claimed to have developed a "novel" insecticide and fungicide derived from a neem seed extract comprising neem oil. This was challenged at the European Patent Office on the grounds that the anti-fungal properties of neem for which the company had obtained the patent had been known in India for centuries, and neem is, in fact, a source of cheap medicine in India. The European Patent Office agreed to proceed with the cancellation of patents.

Basmati rice, which is the most expensive variety of rice and is an important export item, has also suffered because of claims by a US seed company, Rice-Tec, that they have invented novel rice lines whose grains have qualities of Indian basmati rice. The company had already been selling this rice under the names of "Texmati" and "Kasmati". After much research and documentation, the Indian government, through APEDA and the Rice Exporter's Association, successfully filed a "re-examination" application with the US Patent Protection Office, contesting some of the claims in the patent. The company was later forced to withdraw quite a few crucial claims.

As mentioned earlier, these problems largely stemmed from the lack of preparedness on the part of the government to face the new system under the TRIPS regime. The government has now started documenting this traditional knowledge and has also recently enacted "The Geographical Indication of Goods (Registration & Protection) Act". This has become quite an important issue because neither basmati rice nor Darjeeling tea, the two most famous agricultural items, had been accorded protection under the TRIPS Agreement in the WTO. As for basmati rice, the Indian tea exporters' association has also noticed several trademark violations with respect to Darjeeling tea.

REVIEW OF AGRICULTURAL AND FOOD TRADE

Agricultural Trade

Agricultural exports account for a significant share of India's total export basket. During the triennium (hereafter referred to as "TE") ending in 2000, exports of agricultural products accounted for about 12.8 percent share of total exports from India (Table 7). Although the share of agricultural exports in total exports has fallen from 19.5 percent during TE 1988, foreign exchange earnings from agricultural exports have increased significantly. Foreign exchange earnings from agricultural exports increased from US$2.3 billion in TE 1988 to US$3.2 billion in TE 1994, and further to US$4.9 billion in TE 2000. Thus, the total value of agricultural exports exhibited a higher growth during the period between TE 1994 and TE 2000 as than with the growth recorded between TE 1988 and TE 1994. The actual value of agricultural exports in TE 2000 was about 17.7 percent higher than the extrapolated trend value based on 1985-1994 exports.

Imports of agricultural commodities in contrast constitute a relatively small proportion of total imports, and their ratio to agricultural exports has also decreased during the period analysed here. The share of agricultural imports to total imports witnessed a decline from 9.4 percent during TE 1988 to 6.1 percent during TE 1994 but experienced a marginal increase to 7.5 percent during TE 2000. The value of agricultural imports decreased marginally from US$1.6 billion in TE 1988 to US$1.5 billion in TE 1994 and then shot up to US$3.6 billion in TE 2000. Owing to the sudden increase in imports, the actual value of agricultural imports in TE 2000 was 3.1 times the extrapolated trend value based on 1985-1994 imports.

The overall outcome in terms of net agricultural trade was a slight drop in trade surplus from US$1.6 billion in TE 1994 to US$1.4 billion in TE 2000. In comparison with the extrapolated trend value for the period TE 2000 based on 1985-1994 trends in exports and imports, the decrease in trade surplus was quite high, about 54.9 percent.

Net exports showed more or less positive growth from 1989 to 1996, barring the two years between 1992 and 1994 caused by an increase in imports (Figure 1). After 1996, there was a continuous decline in net agricultural exports for three years, 1997, 1998 and 1999, because of the fall in exports and increase in imports.

While the net trade balance of agricultural trade has been favourable all through the study period, there are significant year-to-year variations (Figure 2). During the period from 1989 to 1993, the ratio of agricultural imports to agricultural exports remained below 0.5. In 1994, this ratio jumped to 0.68. It fell below 0.5 in 1995 and remained below this level in 1996 and 1997 as well. This ratio increased significantly in 1998 and 1999 but fell again in 2000. In general, the ratio of agricultural imports to agricultural exports during TE 2000 was about 1.49 times the value of this ratio during TE 1994.

From the data shown in Table 7, both exports and, in particular, imports exhibit a greater stability, which is indicated by the drop in coefficient of variation during the second period as compared with the first period. However, the net trade balance has fluctuated widely during the second period, as reflected in the rise in coefficient of variation.

From these trends in overall exports, it appears that the 1991 economy-wide unilateral reforms, which reduced protection given to the manufacturing sector, introduced the convertibility of the rupee on the trade account and shaped a relatively more open export policy, have had a much greater impact on exports of agricultural products than the changes brought about by the liberalization of agricultural trade under the AoA. This is evident from the trends in the real effective exchange rate (Figure 3) and trends in agricultural exports during 1989-2000. The rising trend in exports was reversed in 1997, and exports continued to decline until 1999. The slowdown in exports during these years is due to several factors such as very little or no trade liberalization, and the significant decline in commodity prices and world trade during 1997, 1998 and 1999. These facts are also supported by the unilateral trade liberalization measures that have been taken during recent years, which are reflected in the trends in individual commodity exports of certain items discussed in the next section.

Exports

The diverse trends in exports of various agricultural items witnessed during the study period have changed the export basket composition of agricultural products somewhat. For example, tea, which accounted for about 19 percent share of total agricultural exports during TE 1988, accounted for only 9 percent share of agricultural exports during TE 2000. This is despite the fact that the actual quantum of exports during TE 2000 was about 40 percent higher than the extrapolated volume for the same period based on the 1985-1994 trend (Table 8). Very similar changes were witnessed in the case of coffee as well.

Likewise, exports of oilseed cakes, which witnessed a very high growth during the period between TE 1988 and TE 1994 (188 percent in quantum terms and 202 percent in value terms) exhibited a significant decline during the second period between TE 1994 and TE 2000. The actual level of exports of oilseed cakes in TE 2000 was 75 percent below the extrapolated value for the same period based on the 1985-1994 trend line. As a result, the share of oilseed cakes in total exports, which had increased from 9.2 percent in TE 1988 to about 20 percent in TE 1994, fell to 8.8 percent in TE 2000. This change in exports of oil cakes is due to unilateral trade policy changes. During the 1980s and the pre-AoA period, imports of edible oils were restricted, which provided higher incentives for domestic production of oilseeds. When imports of edible oils were liberalized in 1994, the output of oilseeds witnessed a slowdown after 1994, which had an impact on the production as well as exports of oil cakes.

However, exports of rice, which accounted for a 9.3 percent share of total agricultural exports during TE 1988, improved its share to 19.5 percent during TE 2000. The level of rice exports increased by 126 percent between TE 1988 and TE 1994 and further by 275 percent between TE 1994 and TE 2000. The actual level of rice exports in TE 2000 was 70 percent higher than the extrapolated trend value. The fall in the unit value of rice exports is due to the change in the product mix of exports of rice.

There are two main types of rice: basmati and non-basmati. The exports of nonbasmati rice were canalized through the Food Corporation of India. In early 1992, the liberalization of non-basmati rice exports was started by permitting exports of some superfine long-grain varieties, subject to a Minimum Export Price. In October 1994, export controls were lifted from all types of rice, including nonbasmati common rice varieties.

These reforms paid rich dividends, which is evident from the remarkable increase in exports of rice. India emerged as the second-largest exporter of rice in 1995, after Thailand, which is the largest exporter of rice in the world (Sharma, 2001). Export of rice from India amounted to 4.9 million tonnes in 1995, comprising 4.5 million tonnes of non-basmati and 0.4 million tonnes of basmati rice, respectively. Rice exports were around 2.5 million tonnes between 1996 and 1997. In 1998 again, the quantity of rice exported, both basmati and non-basmati, touched 5 million tonnes.

Exports of fruits and vegetables (excluding pulses), which account for about 16 percent share of agricultural exports, did not show evidence of any great changes. The average level of exports of fruits and vegetables in TE 2000 was only 23 percent higher than the level of exports in TE 1994 but was 7 percent below the extrapolated trend value for the period TE 2000. Almost similar changes were observed in the case of cashew exports. The exports of cashew nuts in TE 2000 were just 20 percent above the actual level of exports in TE 1994 but were 21 percent below the extrapolated trend value for the period TE 2000.

In the case of spices, exports during the second period between TE 1994 and TE 2000 witnessed a higher growth than that between TE 1988 and TE 1994. This is reflected in the higher increase in the actual level of exports during TE 2000 over the TE 1994 value (48 percent) as compared with the lower growth recorded between TE 1988 and TE 1994 (30 percent). The increase in exports of other items such as meat and meat preparations and tobacco witnessed during the post-AoA period is also promising, but their share in agricultural exports remains low. In the case of cotton, another item in which the country has a comparative advantage, exports have traditionally been subject to quotas announced by the Commissioner of Textiles. Sometimes, exports are even forbidden if there are shortages in the domestic market. As a consequence of these controls, the cotton industry experiences considerable uncertainty as well as periodic shocks. This is reflected in the rise in the coefficient of variation during the second period.

Thus, there are two main reasons for the rather subdued performance of Indian agricultural exports during the study period: first, constraints in the domestic market such as export restrictions and lack of consistency in export policy, which imparted an element of adhocism; second, the international trade environment, which remains quite restrictive and which is now discussed.

First, an examination of the direction of exports of six major agricultural items shows that there are not many changes between TE 1994-1995 and TE 2000-2001, particularly in exports to developed country markets, barring a few exceptions. Second, there are problems of tariff peaks and tariff escalation, particularly in the case of non-traditional exports such as dairy products, fruits and vegetables, preparations of fruits and vegetables, meat and food industry products, which are emerging as promising products. Various studies show that despite national tariff reforms in most countries of the world, high rates of tariffs are still widespread (Table 9).

Third, difficulties have been experienced with the administration of TRQs. It has been observed that minimum access opportunities were provided at the aggregate level, and these are normally distributed in such a way that they hurt domestic producers the least. In addition, there are problems in the allocation of TRQs because most of the time, these are allotted to traditional trading partners, effectively prohibiting the entry of products from other countries. The problems related to TRQs are posing great difficulties in exporting some of the new upcoming items such as mushrooms to the developed countries.

These problems become more complicated because of commodity-specific preferential access arrangements, which are leftovers of past colonial history (Josling, 1998). These bilateral arrangements hinder exports from countries that are not part of the arrangement and are efficient producers. For example, in the case of floriculture, a number of countries such as Gambia, Kenya, Zambia and Zimbabwe benefit from preferences in the EU under the Lomé Convention (now the Cotonou Agreement). Likewise, exports from Colombia and Equador have preferential access in the EU. India had special preferential access for 10 000 tonnes of sugar in the EU market, but the 1995 agreement between the EU and India expired on 30 June 2001.

Fourth, high export subsidies in developed countries continue to be a major constraint for efficient producers despite agreed reductions in export subsidies. Exports which are highly subsidized include dairy products, beef (including other meat), wheat, coarse cereals and sugar, which account for a little over 87 percent share of the total value of export subsidies for all commodities.

Of this, dairy products alone account for about a 29 percent share. Because of high export subsidies on these products, a handful of countries are able to maintain their competitiveness at the expense of countries, such as India, which are not able to use export subsidies but are efficient producers of a wide variety of agricultural commodities. Indian exports that have been seriously impacted include cereals (wheat), dairy products and, to a certain extent, sugar.

Fifth, domestic policies of the developed countries, which allow them to pay huge subsidies to their agriculture, help in reducing the fixed costs of farmers in developed countries. Further, since the basis of the AMS reduction is not commodity-specific, this allows countries to actually increase support for some products while still meeting reduction commitments. The majority of the products that are heavily subsidized by the developed countries directly compete with products that are of interest to countries such as India.

Sixth, there is a large amount of discretion in the SPS Agreement. The aim of the agreement is to prevent member countries from using human, animal and plant health standards for protectionist purposes. Every country, however, has its own rules regarding these restrictions such as inspection of imported products, specific treatment or processing of products, fixing of maximum allowable levels of pesticide residues or permitted use of certain specific additives in food. These flexibilities in the SPS agreement leave considerable room for discretion. In addition, SPS standards are becoming increasingly complex. In the majority of developing countries, the required technology to carry out basic testing and certification is not available. The importance of such standards, and the difficulties that India has experienced during the implementation period, has already been discussed in the section on SPS standards.

Food Trade

Imports of food products account for about 78 percent share of the country's total agricultural imports, but exports of food products account for a lower share, around 56 percent, of total agricultural exports. As food trade accounts for the greater share of both agricultural exports and imports, the behaviour of food trade during the period of study is similar to the trends observed in agricultural trade (Table 10). The value of food imports fell slightly from US$1.4 billion in TE

1988 to US$1.1 billion in TE 1994, and then increased significantly to US$2.8 billion in TE 2000. Owing to the rapid increase in food imports during the period between TE 1994 and TE 2000, the actual value of food imports in TE 2000 was 3.5 times the extrapolated trend value based on 1985-1994 imports.

The value of food exports increased from about US$ 1 billion in TE 1988 to US$ 1.4 billion in TE 1994, and further to US$ 2.7 billion in TE 2000. Hence, as in the case of food imports, the exports of food also exhibited a higher growth during the period between TE 1994 and TE 2000 as compared with the growth recorded between TE 1988 and TE 1994. As a result of the higher growth in food exports during the post-AoA period, the actual value of food exports in TE 2000 was 48 percent higher than the extrapolated trend value based on 1985-1994 exports.

Table 10. Food exports, imports and net food exports

Period	Exports (US$ million)	Imports (US$ million)	Net exports (US$ million)
TE 1988	977	1 356	-378
TE 1994	1 410	1 054	355
TE 2000	2 747	2 770	-23
TE 2000 (E)	1 855	614	1 242
Percentage change			
TE 1994 over TE 1988	44.3	-22.2	194.0
TE 2000 over TE 1994	94.9	162.7	-106.5
TE 2000 over TE 2000 (E)	48.1	351.5	-101.9
Coefficient of variation			
1989-1994	16.29	44.40	81.82
1995-2000	7.92	31.90	149.20

Source: Computed from FAOSTAT database.

Note: E is the extrapolated value based on the log linear trend for the period 1985-1994.

As a consequence of relatively higher growth in food imports than with food exports, net food trade witnessed a reversal from a situation of surplus in TE 1994 of about US$355 million to a deficit of about US$23 million in TE 2000. The trends in yearly changes in net food exports indicate a significant increase in fluctuations, which is evident from Figure 4 and also in the considerable rise in the coefficient of variation during the period between 1995 and 2000 over the period between 1989 and 1994.

On average, although the net balance in food trade remained favourable, there were three years during this period, 1994, 1998 and 1999, when the ratio of food imports to food exports was above unity implying an excess of food imports over food exports. The increase in food imports was largely due to the rise in imports of edible oils during 1998 and 1999.

Edible oils are the only major import item, accounting for a little over 42 percent and 62 percent share of total agricultural imports during the periods TE 1988 and TE 2000, respectively. During the mid-1990s, the share of edible oils in imports declined to 13 percent. These dramatic changes in the imports of edible oils are due to unilateral trade policy reforms implemented in the oilseeds sector. In the 1980s and early 1990s, imports of edible oils remained canalized through the State Trading Corporation (STC) and Hindustan Vegetable Oils Corporation. Imports (mainly of palm oil) were managed so as to support domestic edible oil prices at levels far above world prices to increase oilseed prices and promote oilseed production in the interests of import substitution and self-sufficiency.

Imports of pulses in TE 2000 were 26 percent lower than the level of imports in TE 1994. The extent of the decrease in imports of pulses was close to the level predicted by the extrapolated trend value. The share of imports of pulses has also decreased from 16.3 percent in TE 1994 to 4.7 percent in TE 2000.

In March 1994, imports of palm oil were put under Open General Licence (OGL) at 65 percent import duty, while, at the same time, state agencies like STC and National Dairy Development Board were allowed to import palm oil at a lower duty of 20 percent. In February 1995, this initial liberalization was followed by a far more sweeping reform by which all edible oils except for coconut oil were put under OGL at an import duty of 30 percent. And in May 1999, the import of all types of oilseeds was liberalized and put under OGL at an import duty of 40 percent. Following these far-reaching changes in the edible oils sector, frequent changes have been made in the import duties from time to time, but imports have remained free (Sharma, 2001). These reforms in the edible oil sector and moderate import duties mark a clear break with past policies in the oilseed sector and explain the decrease in imports of edible oils during the six year period preceding the AoA and the sudden increase in imports thereafter.

Among cereals and cereal preparations, wheat is the main import item, imports of which basically reflect occasional shortages. Imports of wheat accounted for about 8.6 percent share of the total value of food imports in TE 1988. Despite significantly higher growth in wheat imports during the period between TE 1994 and TE 2000, they accounted for only 5.6 percent share of food imports during TE 2000. In the 1960s and 1970s, wheat was a major item in the cereal imports of India. The dependence on imports diminished over time, as output levels grew and the country moved towards almost near self-sufficiency in cereals.

As regards imports of sugar, the trend in imports basically reveals a situation of sporadic scarcity. There has not been a major change in the actual level of sugar imports between TE 1988 and TE 1994 or between TE 1994 and TE 2000. But, in comparison with the extrapolated trend value for the period TE 2000, actual imports are much higher. One of the major changes that occurred in the sugar sector after 1994 is the liberalization of sugar imports. Imports initially attracted a very low level of import duty, which has been revised several times and now stands at 60 percent.

Imports of fruits and vegetables also remained restricted because of a general ban on consumer goods. There was no change in this policy until September 1998. When the prices of a few selected vegetables, such as onions and potato, rose unusually sharply, the import of onions and potatoes was put under OGL. As a part of the liberalization process, during the last three years (1999, 2000 and 2001), imports of a number of fruits and vegetables were put under OGL. The trends in imports of fruits and vegetables do not reveal significantly different changes between the period TE 1988-TE 1994 and TE 1994-TE 2000, particularly in quantum terms, although the changes are more significant in value terms mainly because of changes in the unit values of imports.

The post-AoA period has also witnessed changes in the import policy of other products. Dairy products such as skimmed milk powder and butter oil were also freed in 1995, but their levels of imports remain very low.

Livelihood and Food Security Issues

The two main reasons why livelihood and food security issues remain critical for a country like India are (1) the heavy dependence of the majority of the population on the agricultural sector for employment and as a source of income and (2) the large number of undernourished people, the majority of whom are below the poverty line and spend a considerable proportion of their total income on food. About 72 percent of the population of the country lives in rural areas (Census Report, 2001) and the agricultural sector provides employment to about 60 percent of the country's total labour force (GOI, 2002).

A recent FAO study shows that about 225 million people (23 percent of the total population) are undernourished, and about 260 million people (24 percent of the total population) are below the poverty line (GOI, 2002). Although the number of people who are below the poverty line has reduced significantly between 1993-1994 and 1999-2000 (Table 12), it is striking that the number of undernourished people has increased during the period between TE 1992 and TE 1999 (Table 13). The poverty numbers are based on the national data that are generated by the National Sample Survey Organization, and these numbers take into account the actual changes that have occurred in the distribution of consumer expenditure between 1993-1994 and 1999-2000. The food insecurity numbers from the FAO are based on the old data set and do not capture these changes.

Changes in the country's key food security indicators that have occurred during recent years are summarized in Table 14. Per capita food production increased at a much slower pace between TE 1994 and TE 2000, at roughly half the level of growth exhibited during the period from TE 1988 to TE 2000. The supply of food for direct human consumption increased by only 2 percentage points between TE 1994 and TE 2000. This increase in the supply of food was lower in relation to both the extrapolated trend value for the period TE 2000 as well as the level of growth witnessed between TE 1988 and TE 1994.

The other indicators such as per capita food imports and food imports as a ratio of total merchandise exports have also shown significant increases between TE 1994 and TE 2000, after having posted a negative growth between TE

1988 and TE 1994. The growth in the real prices of food products was also more than double the growth observed between TE 1988 and TE 1994.

To a certain extent, some of these changes in imports of food products and food prices were expected after the liberalization of trade in agriculture during the 1990s. But what is more worrying is that there are now signs of declining productivity growth, which, if not addressed, may have adverse implications for the future (Table 15). For these reasons, food self-sufficiency, price stabilization and employment programmes continue to remain the most important objectives of agricultural development policies.

Issues of Concern and Implications for Agricultural Development and Food Security

From this discussion of India's experience during the implementation period, it becomes clear that there are some very serious issues for India arising out of the agreement.

First, since India does not subsidize its agriculture, it is in India's interest that domestic support in those countries that subsidize their agriculture heavily is reduced significantly to the levels that are specified under the agreement as *de minimis* levels. Export subsidies should be eliminated completely. This is because both domestic support and export subsidies, taken together, make exports from countries such as India uncompetitive and affect them adversely.

It is quite evident that reductions in export subsidies and subsidized exports by the developed countries will have greater implications for developing country exports. Because of export subsidies, countries that do not have a comparative advantage are able to maintain their competitiveness at the expense of those that are not able to use export subsidies. It is also important to ensure that greater use of export credits does not substitute for the reduction in export subsidies.

However, as infrastructure facilities in developing countries are generally inadequate, exporters of agricultural commodities in these countries have to bear a very heavy burden of transportation and marketing costs. Subsidies for developing countries under these heads are permitted, and any future attempts to constrain the use of such subsidies would not be in the interest of India.

Second, tariffs peaks, tariff escalation, commodity-specific bilateral agreements and special exemptions remain major issues restricting market access abroad for products in which the country has a comparative advantage. The major product groups where tariff peaks are a problem include dairy products, fruits and vegetables, preparations of fruits and vegetables, meat and fish and food industry products. More meaningful expansion of developed country markets is central to enhance market access for products that are of interest to India.

Third, ambiguities in various parts of the agreement, which basically originate from a lack of clarity, should be taken up so they do not impose restrictions on supporting Indian agriculture in the future. Some of these are related to the calculation of AMS. They are listed here.

(a) In calculating market price support, external reference prices remain fixed based on the years 1986-1988. Thus, if one has to work out the current level of AMS, the gap between the support price and the external reference price must be calculated on the basis of current support price and the base level external reference price. The problem of a fixed external reference price in the calculation of AMS assumes special significance because supporting tables in the original submissions that the member countries have made to the WTO were expressed in domestic currencies. This has created problems in the calculation of AMS for developing countries including India, where inflation rates are generally high compared with developed countries. For example, in the case of India, between the base year, that is TE 1988-89 and the year 2000-2001, the change in the wholesale price index for all commodities works out to be about 146 percent, and depreciation of the Indian rupee is of the order of 214 percent during the same period, respectively. Since the base period submissions were expressed in domestic currency, that is, the Indian rupee, if these changes in inflation and exchange rate are not accounted for, the current AMS would be positive. These problems in the computation of AMS would not have arisen if there were clarity in the agreement on the provisions under which the current total AMS can be adjusted for inflation. The only reference to inflation in the agreement is in Article 18:4, but there is no clarity as to what it implies.

(b) There is insufficient clarity in the agreement whether the quantity eligible to receive the administered price is total production, or only the marketed surplus which is actually sold in the market, or the quantity which is actually procured by the government through the price support mechanisms. Some member countries such as Pakistan have used quantity procured, whereas other countries have used total production. The logic of using total production in these computations is that the government-designated agency is bound to buy whatever is brought to the market at the pre-announced support price. However, there is a limit on this because the quantity brought to the market will not be more than the marketable surplus given that self-consumption accounts for a very large share of the output of basic foodstuffs in a country like India. This is an issue that needs to be resolved in the current negotiations.

The other important question in this context is: How should the value of output be calculated? Should the output be valued at domestic prices or border prices? Since border prices are considered to be the efficiency benchmarks, these should be used to estimate the value of output and not domestic prices as is being practised currently.

(c) There is a lack of clarity in the interpretation of the negative values of product-specific AMS. The problem with this interpretation has significant implications for India, which does not subsidize its agriculture. It is mentioned in Article 7.2(b): "where no Total AMS commitments exist, the member shall not provide support to agricultural producers in excess of the relevant *de minimis* level set out in Article 6.4". This means that countries which do not have reduction commitments are eligible to raise their domestic support up to the *de minimis* level. But Article 13(b)(ii) states that domestic support that conforms to the provisions of the agreement will be nonactionable provided that such measures do not grant support to a specific commodity in excess of that decided during the 1992 marketing year. Because of this ambiguity, it is not clear whether countries could even avail of the *de minimis* level of support allowed in the agreement. This also needs clarification in the ongoing round of negotiations.

In addition, by treating large negative numbers of product-specific support in developing countries as zero, the agreement penalizes countries which tax their agriculture relative to those which subsidize their agriculture. They can continue to subsidize their agricultural production to the tune of 5 percent of the value of agricultural output for both product and non-product-specific support in the case of a developed country or 10 percent for both measures in the case of a developing country.

Apart from this ambiguity, the agreement does allow India to maintain the current system of support to Indian agriculture, because there are no reduction commitments on domestic support as the support provided to Indian agriculture is much below the *de minimis* level which is specified in the agreement. Notwithstanding this flexibility, the agreement does bind the future level of product-specific and non-product-specific support that the government can provide at 10 percent of the value of agricultural output, respectively.

(d) Under Article 6.2, agricultural input subsidies generally available to lowincome or resource-poor farmers in developing countries are exempt from reduction commitments on domestic support. This also applies to domestic support granted to encourage diversification from growing illicit narcotic crops. The problem with this clause is how to define a low-income or resource poor farmer. No guidance is provided in the agreement.

Generally, land-based definitions are used to define resource-poor farmers. However, a land-based definition may not be appropriate if one goes by the actual definition of poverty line in terms of expenditure. Even this will vary from year to year and the poverty line definition used in these computations. Further, definitions based on an average income of US$1 000 for the developing country members referred to in Annex VII of the Agreement on Subsidies and Countervailing Measures would generate a different set of numbers. This is a key issue for India, because exemptions on non-productspecific support to resource-poor farmers allow quite significant room to support agricultural producers.

Fourth, as far as price stabilization is concerned, India has used a three-pronged strategy: price support, public stockholding of cereals and restrictions on external trade. The calculations for domestic support, which were delineated above, showed that India has sufficient flexibility under the agreement to support prices of agricultural products. But the limitations of such policies are already evident from the excessive build up of stocks of cereals, which have been piling up in recent years. The AoA does permit public stockholding for food security purposes, but the cost-effectiveness of such programmes in the light of the fall in international prices and massive subsidization by the developed countries have put serious question marks on the value of such policies.

As regards restrictions on imports to stabilize prices, the abolition of QRs implies that from now on, this would have to be achieved through tariffs on imports. The distribution of bound tariffs on agricultural commodities, which was discussed earlier, shows that this is a viable alternative because the majority of commodities have sufficiently high tariff bindings. In the future, this flexibility may be constrained if further reductions in tariffs occur after the current round of negotiations. In that case, instruments such as variable import tariffs and SSGs would be more useful. However, there is a lack of clarity in the agreement regarding the legality of price bands in the AoA because the footnote to Article 4.2 prohibits the use of variable import duties. If this band is considered as an "ordinary customs duty", it is legal, but if it is considered a variable import duty, it is not legal (Sharma, Greenfield and Konandreas, 1999).

The objective of SSG provisions is to allow the use of additional duties over and above bound rates which can be applied if certain conditions relating to import surges or declines in external reference prices are met. Under the safeguard clause (Article 5) of the AoA, the usage of these provisions is available only for those countries that have bound their tariff levels using the tariffication formula and subsequently scheduled it in their commitments. Future negotiations should consider these issues and allow the usage of these instruments for price stabilization purposes. The use of SSGs will become very important for countries such as India.

Fifth, the level of protection currently given to wines and spirits under geographical indications should be extended to other products such as Darjeeling tea and basmati rice. This is a question of market access for agricultural products as product differentiation is an important element of fair competition. It benefits consumers because they are offered more choice with more information about the actual qualities of products. It also benefits producers, who are able to develop quality products and are protected against unfair or deceptive competition in markets abroad.

Sixth, problems that are being faced by exporters under the SPS Agreement are equally important. Though the aim of this agreement is to prevent member countries from using human, animal and plant health standards for protectionist purposes, every country has its own rules regarding these restrictions such as inspection of imported products, specific treatment or processing of products, fixing of maximum allowable levels of pesticide residues or permitted use of certain specific additives in food. To make matters worse, SPS standards are becoming increasingly complex, and even the required technology to do basic testing and certifications is not available on a wide scale. Countries like India will benefit if developed countries would agree to treat the standards specified by the Codex Alimentarius Commission (CODEX) as equivalent.

Seventh, in the case of commitments on technical and financial assistance, there are essentially two types of issues that need to be addressed. One of these issues is that provisions related to such assistance should be legally binding for the developed countries.

The current provisions of the SDT on technical and financial assistance are in terms of "best endeavour" commitments. Because they are not legally binding, no dispute settlement cases can be initiated on the basis of their non-delivery.

Further, the voluntary assistance which should have been extended has not been forthcoming. The second issue concerns funding arrangements. There is a dire need to set up a special institutional mechanism to support liberalization or globalization and to allocate such funds on a permanent basis so that it does not remain as an *ad hoc* arrangement.

This is essential to build institutions in developing countries to implement reform programmes and also to augment resources to meet the high costs of implementing the provisions needed to make domestic rules conform to international rules and to overcome serious supply-side constraints. Eighth, the issue of food security remains critical for a country like India, but must be handled with utmost care. There is a risk that groups with vested interests may misuse this concern in an emotive way to continue the use of subsidies and protection. There is no denying that food security has remained, and will continue to remain, a vital issue for both developed as well as developing countries.

But there is a difference between the legitimate food security concerns of developing countries, where a majority of the population continues to depend on agriculture for its livelihood, infrastructure is very weak and there are very few avenues for employment outside agriculture for a large mass of illiterates, and developed countries. If developed countries are able to protect their high levels of subsidies and continue their use under the guise of non-trade concerns, it will not be in the benefit of developing countries.

Ninth, under the SDT provisions, the broad framework of the agreement should make provisions that provide real and meaningful market access for the commodities that are produced by countries like India and grants greater flexibility in meeting reduction commitments on domestic support and levels of tariffs.

This is essential for sensitive products to address the concerns of the rural population in this sector for the sustenance of their livelihood and employment. Therefore, the new package of SDT provisions should recognize the two main bottlenecks in implementing any reform in a country like India. First, there is a considerable shortage of institutions to implement reform measures and bear the high costs of implementing the changes. Second, there are serious supply side constraints such as poor infrastructure, which create difficulties in building capacities to compete in the international market.

BIBLIOGRAPHY

Angel, Bradley: *Toxic Threat to Indian Lands: A Greenpeace Report*, San Francisco, Greenpeace, 1991.

Bailey, Ronald: *Liberation Biology: The Scientific and Moral Case for the Biotech Revolution*, Amherst, Prometheus Books, 2005.

Barnum, Susan R.: *Biotechnology: An Introduction*, Belmont, Thomson/Brooks/Cole, 2005.

Burch, William R. Jr.: *Human Ecology: An Environmental Approach*, Belmont, Duxbury Press, 1994.

Campbell, R. R., and Wade, J. L.: *Society and Environment: The Coming Collision*, Boston, Allyn and Bacon, 1972.

Chrispeels, Maarten : *Plants, Genes and Crop Biotechnology*, Sudbury MA, Jones and Barlett Publishers, 2003.

Cox, Earl D. : *Beyond Humanity: Cyber Evolution and Future Minds*, Roackland, Charles River Media, 1996.

Daubenmire, R. F.: *Plants and Environment*, New York, Wiley, 1947.

Dixon, Dougal: *After Man-A Zoology of the Future*, New York, St. Martin's Press, 1981.

Dodds, John H.: *Plant Genetic Engineering*, New York Cambridge University Press, 1985.

Edwards-Rees, D.: *The Service of Youth Book*, Wallington, Religious Education Press, 1943.

Gelobter, Michael: *The Distribution of Air Pollution by Income and Race*, Berkeley, University of California, 1989.

Gitman, Lawrence J.: *Principles of Managerial Finance*, MA: Addison Wesley Longman, 2000.

Johnson, Hazel: *Financial Institutions and Markets: A Global Perspective*, NY, McGraw Hill, 1993.

King, R.B., Mihony, G., and Sheldon, J.K.: *Practical Environmental Bioremediation*, Lewis Publishers, Boca Raton, 1992.

Kruvant, W.J.: *People, Energy, and Pollution*, Cambridge, Ballinger Publishing, 1975.

Martin, Justin: *The Man Behind Money*, Perseus Book Group, 2000.

Matthews, J. E.: *Working with Youth Groups*, London, University of London Press, 1966.

Milson, F.: *Growing With the Job*, London, National Assocation of Youth Clubs, 1968.

Murray, David: *Seeds of Concern: The Genetic Manipulation of Plants*, Sydney, University of New South Wales, 2003.

Primrose, S.B.: *Principles of Gene Manipulation*, London, Blackwell Scientific Publications, 1989.

Retzer, W.J.: *Biotechnology Workbook*, Englewood Cliffs, Prentice Hall, 1991.

Schall and Haley: *Introduction to Financial Management*, McGraw-hill, 1988.

Sontheimer, Sally: *Women and the Environment: Crisis and Development in the Third World*, New York, Monthly Review Press, 1991.

Vallette, J.: *The International Trade in Wastes: A Greenpeace Inventory*, Washington, Greenpeace, 1989.

Walden, Richard: *Genetic Transformation in Plants*, England, Open University Press, 1988.

Zupan, J.M.: *The Distribution of Air Quality in the New York Region*, Baltimore, Johns Hopkins University Press, 1973.

INDEX

R

S

T

W